Special Publications 69

GROUNDWATER VULNERABILITY

Chernobyl Nuclear Disaster

Edited By
Boris Faybishenko
Thomas Nicholson

Vyacheslav Shestopalov
Alexander Bohuslavsky
Volodymir Bublias

This work is a co-publication between
the American Geophysical Union and John Wiley & Sons, Inc.

American Geophysical Union

WILEY

CONTENTS

ABSTRACT

The Chernobyl Nuclear Power Plant (NPP) accident that occurred in Ukraine on 26 April 1986 was one of the most devastating in human history. The monograph is devoted to the problem of groundwater vulnerability using as a case study the results of long-term field and modeling investigations of radionuclide transport in soil and groundwater within the Ukrainian part of the Dnieper River basin (Kyiv region of Ukraine). The authors provide a comprehensive review of existing literature on the assessment of groundwater vulnerability and then describe an improved methodology developed based on integration of the methods of hydrogeological zonation and modeling of anomalously fast migration of radioactive contaminants from the land surface toward groundwater. The monograph also includes the evaluation of the effect of preferential and episodic flow on transport of radionuclides toward the aquifers and risk assessment of groundwater vulnerability.

INTRODUCTION

IMPORTANCE OF LESSONS LEARNED FROM ASSESSMENT
OF GROUNDWATER VULNERABILITY AT CHERNOBYL

A severe nuclear accident at the Chernobyl Nuclear Power Plant in Ukraine (formerly the Ukrainian Soviet Socialist Republic) occurred on 26 April, 1986, causing an extreme explosion and fire. The explosion and fire generated significant airborne radioactive contamination which spread over Ukraine and neighboring countries. The radioactive materials created by the accident measurably contaminated a significant fraction of the Northern Hemisphere [*Hohenemser et al.*, 1986; *Baryakhtar*, 2001; *United Nations Development Program (UNDP)*, 2002]. Important lessons have been learned from studying the Chernobyl accident and its subsequent contamination over approximately the last three decades. Volumes of scientific literature have been written on all aspects of the accident and its consequences.

Why is this monograph devoted to the evaluation of groundwater vulnerability resulting from the Chernobyl accident so important? Because of its sharing of first-hand knowledge and insights on groundwater contamination from this severe accident and the highly detailed monitoring and modeling analyses of radionuclide contaminants (e.g., strontium-90, cesium-137) migration. These studies demonstrate the occurrence of anomalously fast migration of radionuclide contaminants via preferential flow from the land surface to the subsurface groundwater bodies and eventually to surface water bodies.

Although it is assumed that the probability of a severe nuclear accident with significant contamination of the subsurface environment is small, the knowledge gained in these reported Chernobyl studies about fast migration pathways and long-term effects of radioactivity and the insights from consequence management are invaluable to current and future nuclear safety programs. For example, this knowledge and insights may prove useful to those facing the clean-up of the damaged reactors and environs at the Fukushima Daiichi site.

Chernobyl has become a "unique experiment in international collaboration" [*UNDP*, 2002; *Faybishenko et al.*, 2003]. This collaboration has become vital in assessing and predicting the consequences of potential nuclear facility accidents. These assessments and predictions can also be used to mitigate environmental consequences from inadvertent nuclear releases. Because direct observations of the actual migration due to potential nuclear accidents will not be possible for many years, if ever, Chernobyl could be a suitable candidate for providing ana-logue information for a wide variety of nuclear waste disposal and environmental issues around the world. The distinct advantages of the Chernobyl studies over other site-specific studies are due to (1) the Chernobyl Exclusion Zone (CEZ) and the entire territory of Ukraine and its surrounding countries, which have been studied extensively since the 1986 accident; (2) the wealth of data that has been gathered and published about different modes of radionuclide transport in the atmosphere, soils and water bodies, and radionuclide uptake through the various biological food chains; and (3) the use of these databases to provide a unique opportunity to build confidence in characterizing, monitoring, and modeling migration pathways of radionuclides in the biosphere.

The concept that some aquifers are more likely than others to become contaminated has led to the use of the terminology "groundwater *vulnerability* to contamination" [*National Research Council (NRC)*, 1993]. This basic concept of groundwater vulnerability has taken on a range of definitions in the technical literature since the 1960s, and a number of terminologies have been given since then, such as *Vrba and Zaporozec* [1994], *Harter and Walker* [2001], *Harter* [2005], *Thywissen* [2006], *and Manyena* [2006]. *Harter and Rollins* [2008] carried out an extensive review of different terminologies of the term groundwater vulnerability. Examples are "Possibility of percolation of contaminants into water table aquifers," "Degree of endangerment of an aquifer," "Sensitivity of groundwater quality to anthropogenic activities," and "Likelihood for contaminants to reach a specified location in the groundwater system." The *NRC* [1993] defines groundwater vulnerability as "The tendency or likelihood for contaminants to reach a specified position in the groundwater system after introduction at some location above the uppermost aquifer." Thus, groundwater vulnerability can provide a measure of how easy or difficult it is for contaminants at the land surface to migrate and reach an underlying groundwater aquifer. It is a measure of the "degree of insulation" of groundwater from the land surface contaminants [*Harter and Walker*, 2001]. This definition of groundwater vulnerability relates to subsurface contamination from nonpoint or distributed sources of pollution on the land surface.

The *NRC* [1993] also stated that assessing groundwater vulnerability is needed to facilitate policy analysis and development at the local and regional levels, pro-vide program management, inform landuse decisions, and provide general educa-tion and awareness of the region's hydrogeological resources, which ultimately warrant the risk assessment and adequate treatment of groundwater.

The main elements that one needs to consider in a vulnerability assessment for a particular application include (1) the site location and environmental setting, (2) soil morphology and biochemistry, (3) geological and hydrogeological processes and properties, and (4) the contaminant-specific geochemical properties and release mechanism. All of these factors determine the contaminant pathways and mass flux rate from the surface to the upper (unconfined) water table aquifer and then to subsequent underlying aquifers. The *NRC* [1993] identified two general types of vulnerability assessments. The first type is needed to address *specific vulnerability* of the site to a specific contaminant or contaminant class due to human activity. The second type is intended to address *intrinsic vulnerability* of groundwater without consideration of the attributes and behavior of specific contaminants.

The concept of groundwater vulnerability implies that vulnerability is not an absolute or directly measurable property of the hydrogeological system but rather a relative value, or indication, of the relative likelihood (or probability) with which contamination will occur in groundwater systems. It is obvious to state that *all groundwater systems are vulnerable, and uncertainty is inherent in all vulnerability assessments* [*NRC*, 1993]. In the context of management of groundwater contaminated with radioactive contaminants, policy makers, resource managers, and land users would use vulnerability assessments as a tool to adjust regulatory requirements and management for different practices as well as to evaluate the risk of using groundwater as a potable water supply source or for irrigation purposes. The results of the vulnerability assessment of groundwater at Chernobyl can provide a very important contribution to the enhancement of the databases, which are currently the source of much uncertainty in vulnerability assessment.

The majority of existing vulnerability assessment techniques address mainly transport due to large-scale infiltration and percolation to groundwater systems. These techniques generally ignore an accelerated (and often episodic) transport through preferential flow pathways in the vadose zone, such as root zones, cracks, joints, fractures, and near-surface weak zones below land surface depressions. Therefore, the vulnerability assessment is highly dependent on the spatial resolution of the field-scale site mapping and modeling, especially preferential flow pathways. Noninclusion of preferential flow pathways in the vulnerability assessment is a significant limitation and constraint in the risk assessment of groundwater contamination.

Overall, the transport through preferential flow zones is a major driver of contaminants from the land surface to groundwater and must be taken into account in the groundwater vulnerability assessment to effectively assess and possibly manage risk due to groundwater contamination. Understanding where and how radionuclides reach the accessible environment requires knowledge of the behavior of shallow groundwater systems such as perched water conditions and their relationship to ephemeral, intermittent, and perennial streamflows. The seasonal nature of groundwater recharge and plant uptake needs to be considered. Assumptions of natural attenuation to always dilute and disperse residual radioactivity should be tested through detailed monitoring of the soil-groundwater systems, their biogeochemical character, and plant uptake processes. In some cases, biogeochemical processes within surface depressions may enhance accumulation of contaminants in the near-surface zone, leading to the elevated concentrations of contaminants in groundwater.

The current monograph synthesizes the knowledge and insights from the highly diverse and detailed field and modeling studies conducted following the

Chernobyl accident. The development of this monograph further extends the research understanding of the Chernobyl consequences, which the authors have carried out in the CEZ since the accident [*Shestopalov et al.*, 2006]. In this monograph, the authors critically assessed the applicability of commonly used methods [*Vrba and Zaporozec*, 1994] to evaluate groundwater vulnerability. These assessments are based on the calculations of various types of indices and parameters describing the vulnerability and protective capability of soil and hydrogeological systems. Unfortunately, most of the existing methods do not take into account the observed anomalously fast migration pathways of radionuclide contaminants through preferential flow and migration zones (PFMZs). The properties of the PFMZs are a function of the land surface, biogeochemistry, soil heterogeneity, and hydrogeological systems at scales ranging from millimeters to kilometers. Based on the results of field experimental and modeling studies, the authors show that while the averaged areal radionuclide penetration to subsurface groundwater aquifers is slow, near-surface flow and radionuclide contaminants converge through local surface depressions and reach the groundwater table or perched systems relatively fast through the subvertical PFMZs.

Chernobyl studies clearly show that the evaluation of the effect of different-scale PFMZs is important not only for calculations of vulnerability of shallow unconfined groundwater but also for deeper confined aquifers, especially those within the tectonic disturbed zones and other linear forms of PFMZs, such as those typical for the Fukushima area and the nuclear waste disposal sites.

The results summarized in this monograph can be used to evaluate groundwater vulnerability and conduct risk assessments for groundwater contamination from diverse sources as well as selection of optimal remediation methods for reducing threats from residual radioactivity, depending on the types of sediments and hydrogeological systems and an aerial distribution of PFMZs.

REFERENCES

Baryakhtar, V. (Ed.) (2001), Chernobyl, Exclusion Zone, Collected Papers, Kyiv, Naukova Dumka.

Cardona, O. D., M. K. van Aalst, J. Birkmann, M. Fordham, G. McGregor, R. Perez, R. S. Pulwarty, E. L .F. Schipper, and B. T. Sinh (2012), Determinants of risk: Exposure and vulnerability, in Managing the Risks of Extreme Events and Disasters to Advance Climate Change Adaptation, edited by C. B. Field et al., Special report of Working Groups I and II of the Intergovernmental Panel on Climate Change (IPCC), pp. 65–108, Cambridge Univ. Press, Cambridge, United Kingdom.

Faybishenko, B., A. L. Young, V. G. Baryakhtar, A. L. Taboas, and L. Habegger (2003), Reflections on the Chernobyl accident and the future of nuclear power, Environ. Sci. Pollut. Res., Special Issue 1, 1–2.

Harter, T. (2005), Vulnerability mapping of groundwater resources, Water Encyclopedia, 561–566, doi:10.1002/047147844X.gw1236.

Harter, T., and L. Rollins (Eds.) (2008), Watersheds, Groundwater and Drinking Water: A Practical Guide, Univ. of Calif. Agric. and Natural Resources Publication 3497, Berkeley, CA.

Harter, T., and L. G. Walker (2001), Assessing vulnerability of groundwater, U.S. Nat. Resources Conservation Service, Washington DC.

Hohenemser, C. M. Deicher, A. Ernst, H. Hofsass, G. Lindner, and E. Recknagel (1986), Chernobyl: An early report, Environment, 28(5), 6–13, 30–43.

Manyena, S. B. (2006), The concept of resilience revisited, Disasters, 30(4), 433–450.

National Research Council (NRC) (1993), Groundwater vulnerability assessment, contamination potential under conditions of uncertainty, Committee on Techniques for Assessing Ground Water Vulnerability, NRC, Nat. Acad. Press, Washington, D.C.

Shestopalov, V. M., Yu. F. Rudenko, A. S. Bohuslavsky, and V. N. Bublias (2006) Chernobyl-born radionuclides: Aquifers protectability with respect to preferential flow zones, in Applied Hydrogeophysics, edited by H. Vereecken, pp. 341–376, Springer, Netherlands.

Thywissen, K. (2006), Core terminology of disaster risk reduction: A comparative glossary, in Measuring Vulnerability to Natural Hazards, edited by J. Birkmann, pp. 448–496, UNU Press, Tokyo, Japan.

United Nations Development Program (UNDP) (2002), The Human Consequences of the Chernobyl Nuclear Accident, A Strategy for Recovery, report commissioned by the United Nations Development Program, http://www.un.org/ha/chernobyl.

Vrba J. and Zaporozec A. (Eds.) (1994), Guidebook on Mapping Groundwater Vulnerability. International Contributions to Hydrogeology, vol. 16, Int. Assoc. of Hydrogeol, Heise, Hanover.

1. METHODS OF GROUNDWATER VULNERABILITY AND PROTECTABILITY ASSESSMENT

Step 1 in an assessment of groundwater vulnerability and protectability from pollution within a certain area includes the evaluation of hydrogeological and physicochemical processes and factors followed by subdivision of the studied area into zones with similar geological and hydrogeological conditions. *Methods of hydrogeological zoning* were used for groundwater vulnerability assessment by *Margat* [1968], *Vrana* [1968, 1984], *Albinet and Margat* [1970], *Rogovskaya* [1976], *Josopait and Schwerdtfeger* [1979], *Ostry et al.* [1987]. Then, there were developed parametric methods and corresponding scoring and *index* system methods that were used to quantify the most significant characteristics of the geological medium (lithology, hydralic conductivity, infiltration, etc.). Among these methods are DRASTIC [*Aller et al.*, 1987; *Rosen*, 1994], SINTACS [*Civita and De Maio*, 2004], GOD [*Foster*, 1987; *Foster and Hirata*, 1988], and other index-rating assessment methods [*Villumsen et al.*, 1983; *Engelen*, 1985; *Zaporozec*, 1985; *Andersen and Gosk*, 1987; *Carter et al.*, 1987; *Marcolongo and Pretto*, 1987; *Schmidt*, 1987; *Sotornikova and Vrba*, 1987; *Palmer*, 1988; *Doerfliger et al.*, 1999; *Magiera*, 2000; *Rogachevskaya*, 2002]. With development of modern geographic information systems (GISs) and mapping techniques (maps overlay, three-dimensional (3D) data processing and visualization, etc.), these methods became more complex and detailed, taking into account an increasing number of hydrogeological, geological, climatic, and other parameters and criteria [*Engel et al.*,1996; *Burkart et al.*, 1999; *Zhou et al.*, 1999; *Gogu and Dassargues*, 2000; *Zaporozec*, 2002; *Daly et al.*, 2002; *Zwahlen*, 2004; *Sinreich, et al.*, 2007; *Ligget and Talwar*, 2009].

On the other hand, in parallel to the development of zoning and index-rating methods, even before the appearance of groundwater vulnerability and protectability concepts in the 1960s, many researches used characteristic unified physicochemical parameters of the geological medium, such as *travel time* necessary for the contamination front to reach groundwater from the contamination source, the *retardation factor* (ratio of velocities of seepage water and contaminant particles), or *hydraulic resistance* of covering and water-bearing deposits. These characteristic

---

Groundwater Vulnerability: Chernobyl Nuclear Disaster, Monograph Number 69.
Edited by Boris Faybishenko and Thomas Nicholson.
© 2015 American Geophysical Union. Published 2015 by John Wiley & Sons, Inc.

parameters became the basis for the groundwater vulnerability and protectability assessment, and the corresponding methods can be called methods of *parametric assessment*. This approach was developed in publications of the former USSR researchers [*Goldberg*, 1983, 1987; *Mironenko and Rumynin*, 1990; *Belousova and Galaktionova*, 1994; *Belousova*, 2001, 2005; *Pityeva*, 1999; *Pashkovskiy*, 2002; *Rogachevskaya*, 2002; *Zektser*, 2007] as well as by Western authors [e.g., *Van Stempvoort et al.*, 1995]. Parametric methods of the assessment of groundwater vulnerability were then combined with numerical modeling techniques to incorporate the effect of complex hydrogeological and geological conditions along with physicochemical parameters characterizing the geological medium and interactions in the "contaminant-water-rock" system [*Rumynin*, 2003; *Loague et al.*, 1998; *Shestopalov et al.*, 2006; *Zhang et al.*, 1996].

In this chapter, the authors will discuss in more detail existing methods of groundwater vulnerability assessment based on hydrogeological zoning, index rating, parametric, and modeling methods.

1.1. Method of Hydrogeological Zoning

Starting from the early works of the 1960s, the methods of groundwater vulnerability assessment and mapping were developed based on subdividing the studied area into a number of zones with different degree of vulnerability based on zoning by hydrogeological conditions, relief, thickness and composition of soil and vadose zone, etc. The groundwater vulnerability is represented by qualitative categorization of groundwater into several "homogeneous" zones, for example, of very low, low, medium, high, and very high vulnerability. The resulted zones with different vulnerability degree in most cases are obtained using the procedure of "overlaying maps" of basic initial data on which the homogeneous zones are first contoured corresponding to different types or degree of the initial (basic) characteristic. This procedure became easier to perform in detail using modern GIS technologies. Strictly speaking, the initial data zoning and categorization procedure is a necessary and useful preliminary stage for any groundwater vulnerability assessment.

The method of area zoning by hydrogeological features was used in the first groundwater vulnerability assessments [*Vrana*, 1968; *Albinet and Margat*, 1970; *Olmer and Rezac*, 1974]. A case study is presented by *Sililo et al.* [2001]. They developed a system of regional qualitative groundwater protectability assessment for South Africa in the scale 1:250,000 using GIS overlay of initial maps in the scale 1:50,000 of relief, climatic characteristics, and type and composition of soil. After performing the zoning procedure on the initial data, they built maps of clay fraction and iron content and obtained resulting maps of attenuation potential of soil separately for cation- and anion-forming groups of contaminants. The maps include three classes of attenuation potential: low, medium, and relatively high.

Another example of the zoning method combined with modern GIS technology is the development of a regional groundwater vulnerability map of Scotland (in scale 1:100,000) by *Ball et al.* [2004] in the framework of the SNIFFER project. They performed the initial data analysis by overlaying maps of soil and vadose zone thickness, lithology and permeability, character of aquifer occurrence, hydraulic conductivity, porosity, and fracturing degree of rocks. As a result they classified the study area into seven characteristic types of hydrogeological conditions determined by most frequently occurring lithological sections of vadoze zone and character of groundwater occurrence: (1) highly permeable alluvium-delluvium deposits (drift), (2) exposed hard fractured rocks, (3) hard fractured rocks covered by soil layer, (4) hard fractured rocks covered by drift layer, (5) fractured open rocks with double porosity, (6) fractured rocks covered by soil layer, and (7) fractured rocks covered by drift layer. According to these types, seven scenarios of groundwater vulnerability categorization have been developed which include 199 different vulnerability codes (possible combinations of gradations for thickness and hydraulic conductivity of layers for the above seven section types). In the study area of Scotland, only 46 of these 199 gradations occur. The resulting map of groundwater vulnerability is obtained after the GIS zoning procedure according to the above seven types of groundwater occurrence. The authors conclude that the majority of the studied area of Scotland has maximum or very high groundwater vulnerability because of wide occurrence of highly fractured weathered hard rocks, often uncovered or covered by thin layers of soil and highly permeable drift.

The hydrogeological zoning method is able to provide broad-scale regional groundwater vulnerability maps, including modern GIS-based maps with high resolution which use large volumes of data of hydrogeological, geological, climatic, relief, and other characteristics. An assessment system and gradations developed using this method in most cases are targeted only to the assessment area for which it was developed, and they cannot be used without special adaptation for groundwater vulnerability assessments of other areas.

1.2. Index Methods

The necessity of fast and effective assessments of the groundwater pollution risks related with increasing requirements of municipal services of water supply, farms, environment protection agencies, etc., in the United States, France, Italy, Germany, and other countries, starting from the 1980s, stimulated the development of different index-type and rating-type assessment systems of groundwater contamination risks, groundwater vulnerability and protectability based on simple algorithms of unification (summation, generalization) of parameters, and factors characterizing the hydrogeological conditions and protection ability of the geological medium containing the assessed groundwater. The appearance of modern GIS technologies allowed for the development of several, effective methods which have already been used in different countries.

The *DRASTIC* method was proposed by *Aller et al.* [1987] for the U.S. Environmental Protection Agency (EPA) and was applied in the United States, Canada, South Africa, and many other countries. The method is based on the calculation (at each point of the assessed area) of a unified groundwater vulnerability index, DRASTIC, as a sum of seven rating indicators (*D, R, A, S, T, I, C*) multiplied by the corresponding weight factors r_1 through r_7:

$$DRASTIC = r_1 \cdot D + r_2 \cdot R + r_3 \cdot A + r_4 \cdot S + r_5 \cdot T + r_6 \cdot I + r_7 \cdot C,$$

where *D* is the groundwater table depth, *R* the net recharge, *A* the aquifer media (determined by lithology), *S* the soil type (by texture), *T* the topography (by slope), *I* the impact of the vadose zone, and *C* the aquifer hydraulic conductivity.

Each indicator is assessed by the corresponding local hydrogeological characteristic in a 10-point scoring system [*Aller et al.*, 1987]. For weight coefficients r_1–r_7, determining the relative "importance" of the corresponding indicator, two sets of values are proposed: (1) r_1–r_7 = 5,4,3,2,1,5,3 and (2) r_1–r_7 = 5,4,3,5,3,4,2. The first set determines the standard DRASTIC index used in most cases for assessing the *intrinsic* groundwater vulnerability, and the second set ("agricultural" DRASTIC) is designed for *special* vulnerability to contamination with pesticides. Thus determined, the assessed DRASTIC index can vary within the range of 23–230 (intrinsic vulnerability) or 26–260 (vulnerability to pesticides). The proposed values of weight coefficients to some degree have the judgmental character and /or are based on experimental results. It is clear from the above formula for the DRASTIC index that, in case of pesticides, the soil type and slope angle appear to be more important, but the influence of the vadose zone and the aquifer conductivity are less important.

Higher values of the DRASTIC index correspond to higher groundwater vulnerability. In real practical applications, the DRASTIC index usually varies in the range of 5–200.

For example, *Zektser et al.* [2004] used DRASTIC (with the second set of weight coefficients) for building a vulnerability map of the main aquifer in Castelporciano province (Italy). They obtained the DRASTIC index in the range 26–256 and determined five groundwater vulnerability categories: 26–72, very low; 72–118, low; 118–164, medium; 164–210, high; and 210–256, very high. *Denny et al.* [2007] proposed to modify the DRASTIC method in order to incorporate the structural characteristics of bedrock aquifers with large-scale fracture zones and faults acting as primary conduits for flow at the regional scale. The methodology is applied to the southern Gulf Islands region of southwestern British Columbia, Canada. Bedrock geology maps, soil maps, structural measurements, mapped lineaments, water well information, and topographic data assembled within a comprehensive GIS database are used to assess the traditional DRASTIC indices, and additional structural indices are considered for accounting the regional structural elements during the recharge and well capture zone determinations.

SINTACS was developed in works by *Civita and De Maio* [2004] and *Civita* [2008] for use in Italy. It represents a more detailed and refined variant of DRASTIC. Similarly to DRASTIC, the SINTACS index is determined as a sum of seven weighted indicators (ratings): S (*soggicenza*), depth to groundwater (range 0–100 m); I (*infiltrazione*), recharge (0–550 mm/year); N (*non saturo*), vadose zone lithology with account of fracturing; T (*tipologia della copertura*), soil type (composition); A (*acquifero*), saturated zone (aquifer) characteristic (composition, disturbance, including karst occurrence); C (*conducibilità*), hydraulic conductivity; and S (*superficie topografica*), topography (slope).

In contrast to DRASTIC, the table of scores for each SINTACS indicator contains more detailed lithological differences and disturbances (fractures, karst). Authors have developed the five series of weight coefficients, r_1–r_7, according to types of hydrogeological conditions of the study area and an additional set for assessment of the special groundwater vulnerability to nitrate contamination:

1. Normal recharge: r_1–r_7 = 5,4,5,3,3,3,3
2. High (technogenic recharge): r_1–r_7 = 5,5,4,5,3,2,2
3. Temporarily flooded areas, with account of watercourse density: r_1–r_7 = 4,4,4,2,5,5,2
4. Karst rocks: r_1–r_7 = 2,5,1,3,5,5,5
5. Fractured rocks: r_1–r_7 = 3,3,3,4,4,5,4
6. Nitrate contamination: r_1–r_7 = 5,5,4,5,2,2,3

An important feature of the method is its attempt to account implicitly for hydraulic conditions of the vadose and saturated zones, including types of rocks, technogenic recharge, flooding, etc.

GOD is an index-rating method of assessing regional groundwater vulnerability proposed by *Foster* [1987, 1988] for geological conditions of Great Britain where the groundwater occurs mainly in fractured rocks (limestones, sandstones) overlaid with unconsolidated deposits of the vadose zone and soil. The method is based on the evaluation of three groundwater vulnerability indicators:

1. Types of an aquifer — unconfined, confined, or confined-unconfined groundwater
2. Overall lithology — composition of covering deposits, aquifer rocks, degree of consolidation
3. Depth to groundwater

The authors of this method stress attention to accounting for fracturing and other rock heterogeneities. Each of three indicators ranges in value from 0 (minimum vulnerability) to 1 (maximum vulnerability). The resulting GOD index is determined as the product of all three indicators, and the groundwater vulnerability map is obtained as a distribution of the GOD index over the studied area.

SUPRA is a regional mapping method for groundwater vulnerability. It is an index method using the matrix ranging procedure for the indicators and GIS overlay to obtain the resulting areal groundwater vulnerability. The method is proposed by Zaporozec [2002] and was applied in mapping the groundwater vulnerability of northern Wisconsin. The assessment is based on five indicators:

1. Soil characteristics
2. Unsaturated zone thickness
3. Permeability of vertical sequences in the unsaturated zone
4. Groundwater recharge
5. Aquifer characteristics (lithology, flow regime, recharge)

The resulting vulnerability index is assessed in three stages that correspond to the assessment objectives and stages of downward migration of contaminants from the soil surface into groundwater aquifers:

 I. Evaluation of the soil capacity to attenuate contaminants
 II. Evaluation of the contamination potential of shallow groundwater
III. Evaluation of the contamination potential of deeper aquifers

In the conclusion of each stage, a map is built which can be used alone or in compiling the combined composite map. Use of the independently evaluated components (soil, upper aquifer, deeper aquifers) makes the method flexible to requirements of different users.

The assessment and mapping of vulnerability of deeper aquifers (stage III) is based on geological and hydrogeological characteristics such as aquifer deposit lithology, integrity of the overlying confining bed, location, area and character of the recharge zones, as well as the regional groundwater flow direction.

The evaluation of attenuation capacity of soil can be carried out using the soil contamination attenuation model (SCAM) developed by *Zaporozec* [1985]. With its aid the soil attenuation capacity is assessed in relation to contaminant sources, located within or out of the soil, based on a two-layer model (soil and subsoil), using characteristic indicators such as the soil texture, pH, depth, drainage degree, and content of organic material. Each indicator is assessed by its score, and the sum of the scores is found using GIS. Depending on the total score range in the area, it is classified based on the four categories of soil attenuation capacity: best, good, average, and least.

The second assessment stage for the upper groundwater aquifer is performed using GIS based on the three parameters: unsaturated zone thickness, vertical hydraulic conductivity, and average groundwater recharge assessed by means of the evaluation of net infiltration. Each of these parameters is assessed using three gradations (low, medium, and high). After that, for each assessed subarea, the GIS matrix overlaying procedure is performed successively for these three indicators, in the result of which the groundwater vulnerability is assessed as low, medium, or high, and a three-color map (green, yellow, red) of the corresponding vulnerability zones is drawn for the studied area.

As is noted by Zaporozec, the method is designed as a base for general land use and construction planning.

The *DRAW method* is described by *Zhou et al.* [2010]. The method was developed in China for groundwater vulnerability assessments in arid areas. For calculating the overall vulnerability index, the method combines four main assessment characteristics: D, the depth; R, the net recharge of the aquifer; A, the aquifer characteristics; and V, the lithology of the vadose zone. As a case study, the *Zhou et al.* [2010] paper assesses the vulnerability of a phreatic aquifer in Tarim Basin of Xinjiang. As reported by the authors, the groundwater vulnerability zones with vulnerability index ranging within 2–4, 4–6, 6–8, and >8 account for 10.1, 80.4, 9.2, and 0.2%, respectively, of the total plain area of the Tarim Basin. The areas with the latter two higher vulnerability ranges (6–8 and >8) are mainly located in the irrigation districts with thin soil layer (20–30 cm thick near-surface soil of vadose zone, mainly with underlying sandy gravel) and with silty and fine sand layer. Such a vadose zone generally lacks low permeability sandy loam and clayey soil, resulting in greater recharge due to infiltration of irrigation water.

The *EPIK* method was designed specially for use at karst areas in Switzerland by *Doerfliger et al.* [1999] for assessment of groundwater vulnerability of karstic alpine areas. The method is based on the classification of lithology and permeability of the unsaturated zone, recharge conditions, and karst development. The following four indicators are used:
1. Epikarst (weathered fractured bedrock layer beneath the soil or at the surface)
2. Protective cover
3. Infiltration conditions (with account of relief)
4. Karst development

The scores of these indicators are summed and weighted using the expert evaluation. The final assessment gives three categories of groundwater vulnerability: average, high, and very high. The method was used to assess the influence of karst on the groundwater vulnerability at test sites in Switzerland, Spain, and Germany during the implementation of the COST-620 Project [*Zwahlen*, 2004].

The *German State Geological Survey (GLA)* method was developed by *Hoelting et al.* [1995] for regional protectability assessment of the upper groundwater aquifer. The method accounts for the protective effectiveness of soil (down to a depth of 1 m, the average rooting depth) and the unsaturated zone. The assessment is based on the scores of the following indicators:
1. Effective moisture capacity of soil, S (mm), takes scores 10 (0–49 mm), 50 (50–89 mm), 125 (90–139 mm), 250 (140–199 mm), 500 (200–249 mm), and 750 (≥250 mm).
2. Percolation rate, W (mm/year) takes scores from 2.25 to 0.5 for increasing groundwater recharge from 0 to over 400 mm/year.
3. Type of rock is given as $R = O \cdot F$, where O and F are defined as follows:
   O is the rock type with scores 5 (conglomerate, breccia, limestone, dolomite, etc.); 10 (porous sandstone, porous tuff); 15 (sandstone,

quartzite, massive igneous and metamorphic rock); 20 (claystone, silt-stone, shale, marlstone);

F is the jointing and karstification indicator with values: 0.3 (strongly jointed, fractured, or karstic), 0.5 (moderately karstic), 1(moderately jointed, slightly karstic; or no data), 4 (slightly jointed), and 25 (nonjointed).

4. Unsaturated zone thickness T (sum of layer thicknesses T_n).

The resulting groundwater protectability index P_T is calculated as

$$P_T = P_1 + P_2 + Q + HP,$$

where

$$P_1 = S \cdot W$$

is the protective effectiveness of the soil;

$$P_2 = W \left(R_1 T_1 + R_2 T_2 + \cdots + R_n T_n \right)$$

is the total protective effectiveness of the unsaturated zone layers, accounting for lithology and rock disturbance (fracturing, karst); and $Q = 500$ and $HP = 1500$ are scores added a perched aquifer and a confined aquifer, respectively, if present. Using the P_T index, the groundwater protectability categories are determined as follows: very low ($P_T < 500$), low (500–1000), average (1000–2000), high (2000–3000), and very high (3000–4000).

The most problematic procedure of the method is the selection and substantiation of the separate indicator scores for a given area.

The method was used in Germany [*Von Hoyer and Söfner*, 1998] and other countries [*Margane et al.*, 1999]. It became the base for the PI method described below.

The *PI method* of regional assessment of intrinsic groundwater vulnerability was developed by *Goldscheider* [2005] especially for karst areas, but it can be used for any other hydrogeological conditions. Although based on the German method described above, in contrast, it specifically takes into account the zones of fast infiltration related with the accumulation of surface runoff in depressions and direct influx of water into the upper aquifer through the open karst forms (caves, holes, etc.).

The resulting index of groundwater vulnerability is assessed as a product of two indicators:

1. Protective capacity of soil and unsaturated zone, P, with scores 1 (very low), 2 (low), 3 (average), 4 (high), and 5 (very high). Increasing scores by one point corresponds to a tenfold increase of protective capacity.
2. Infiltration conditions, I, is an indicator of the influence of fast infiltration zones, with scores 0–0.2 (maximum), 0.2–0.4 (high), 0.4–0.6 (average), 0.6–0.8 (low), and 0.8–1 (very low).

The protective capacity is assessed by score tables accounting for lithology, effective capacity, granulometry, fracturing, and karst. The detailed score tables are given in the COST-620 Project report [*Zwahlen*, 2004].

The final gradations of groundwater vulnerability PI index are as follows: PI = 4–5 (very low vulnerability), 3–4 (low), 2–3 (average), 1–2 (high), and 0–1 (maximum). In fact, this index gives the groundwater *protectability* rather than vulnerability, as it increases with the decrease in vulnerability (increase in protectability) of the assessed groundwater.

The PI method is used in the "European approach" to groundwater vulnerability assessment of karstic areas [*Daly et al.*, 2002] developed during the European Community (EC) COST-620 Project. In the result of this project, the *COP method* was designed based on three main indicators: (1) concentration of flow, (2) overlying layers, and (3) precipitation regime. The first of these indicators (*C* and *O*) correspond to the *I* and *P* indicators, respectively, of the PI method described above, and the third one (*P*) is a climatic indicator accounting for the annual atmospheric precipitation, frequency, duration, and intensity of precipitation events [*Zwahlen*, 2004].

The modified European approach was developed by *Shestopalov et al.* [2009] in Ukraine (called by authors "the Mountain Crimea approach") for assessment of karst groundwater vulnerability. In this approach the COP method was adapted and modified for specific conditions of the area of Ai-Petri karst massif in mountainous Crimea representing the main recharge area of the regional groundwater system. The modification of the European approach includes accounting for the special properties of the epikarst and concentration of the underground runoff by karst caves. The GIS-based resulting map of assessed groundwater vulnerability in the PI method scale is obtained for the research area.

From the above consideration, it can be concluded that the common feature of the index-rating methods is in a significant degree "judgmental" approach to the definition of rating scores and scales for main factors and indicators of groundwater vulnerability.

1.3. Parametric Methods

The most known system of *groundwater protectability assessment* standardized in the former USSR was developed by *Goldberg* [1983, 1987], who determined the *groundwater protectability* to be the state of overlaying of an aquifer by deposits, first of all low-permeable ones, which prevent the penetration of contaminants from the land surface into groundwater. According to his representation, the groundwater's protectability depends on a number of factors which can be classified into three main groups, natural, technogenic, and physicochemical, as follows:

1. *Natural* Factors   Presence of low-permeable deposits in the vertical section; depth to groundwater table; thickness, lithology, and permeability properties of rocks (first low-permeable) overlying the aquifer; capacity (sorption)

properties of rocks; and interrelation of groundwater heads (levels) in the studied and overlying aquifers.

2. *Technogenic* Factors   Presence of contaminants on the land surface (waste collectors, slurry tanks, pits, outflow of wastewater over the watershed areas, irrigation with wastewater, etc.) and character of contaminant influx to groundwater determined by these conditions.

3. *Physicochemical*   Specific properties of contaminants, their migration, sorption, and degradation properties chemical stability, and interaction with groundwater and rocks.

According to *Goldberg* [1983], the complete groundwater protectability assessment requires all the above factors to be taken into account. As a complex characteristic determining the risk of groundwater contaminatnion, Goldberg introduces the groundwater *susceptibility Π* to contamination as determined by the ratio

$$\Pi = M_T/\varepsilon, \tag{1.1}$$

where M_T is the module of the technogenic load assessed in thousands of tons of contaminant fallout per square kilometer of land surface in a year, and ε is the dimensionless *groundwater protectability index* assessed in relative units.

From the above three main groundwater protectability factors, the natural ones are of primary importance because they determine the degree of the natural protection of an aquifer from any contaminants and conditions of their penetration from the land surface. Among the natural factors, the most important is the presence of overlying low-permeable deposits: clays, heavy loams, loams, sandy loams, and loamy sands with hydraulic conductivity k below 0.1 m/day.

Quantitatively the groundwater protectability can be characterized by the dimensionless index ε as described below.

As the main parameter of groundwater protectability, Goldberg used the *percolation time t_w* that is the time needed for percolating contaminated water from the land surface to reach the groundwater table. For the upper (unconfined) aquifer the time t_w is assessed for the following two scenarios:

1. Flow of contaminated (waste) waters from the surface basins with constant level H_c. The percolation time is determined by the formula

$$t_w = (n - n_e) H_c / k \left[(m/H_c) - \ln(1 + m/H_c) \right], \tag{1.2}$$

where H_c is the height of the wastewater column in the basin (the average value $H_c = 5$ m is usually taken in groundwater protectability assessments), k the unsaturated zone hydraulic conductivity (m/day), m the unsaturated zone thickness (m), n the porosity, and n_e the initial soil moisture content in the unsaturated zone.

2. Flow of contaminated water with constant flow rate Q_c (m³/day) with corresponding percolation rate $w_c = Q_c/F$, where F (m²) is the recharge surface area. In the case of $w_c \leq k$, where k (m/day) is the hydraulic conductivity of the unsaturated zone, the percolation time is determined by the formula

$$t_w = \frac{mn}{\sqrt[3]{w_c^2 k}} \tag{1.3}$$

For $w_c > k$ (a temporary layer of contaminated water is formed on the surface), the time t_w is determined by the formula

$$t_w = \frac{m}{\dfrac{(1-n)k}{2n} + \sqrt{\dfrac{(1-n^2)k^2}{4n^2} + \dfrac{qk}{n}}}. \tag{1.4}$$

In a heterogeneous stratified unsaturated zone, the equivalent hydraulic conductivity of the averaged section can be determined by the formula

$$k_e = \frac{m}{m_1/k_1 + m_2/k_2 + \cdots + m_i/k_i}, \tag{1.5}$$

where $m_1, m_2, \ldots, m_i$ and $k_1, k_2, \ldots, k_i$ are the thicknesses and hydraulic conductivities, respectively, of the layers.

The groundwater protectability index ε for the *upper* (*unconfined*) *groundwater* is assessed using Goldberg's qualitative groundwater protectability assessment by an integer sum of two scores: (1) for the depth to groundwater table, H, and (2) for the low-permeable layers in the unsaturated zone (if present). The first one takes values 1–5 for corresponding intervals of groundwater table depth, as determined in Table 1.1.

If low-permeable deposits ($k \leq 0.1$) are present in the unsaturated zone then the second (additional) score is determined by the total thickness m_0 and hydraulic conductivity k as given in Table 1.2 for different lithology groups. The resulting groundwater protectability index ε is assessed by finding the sum of the scores (Tables 1.1 and 1.2) ranging from 1 to 30; according to this range, Goldberg determined six groundwater protectability categories:

Category I: $\varepsilon \leq 5$
Category II: $5 < \varepsilon \leq 10$
Category III: $10 < \varepsilon \leq 15$
Category IV: $15 < \varepsilon \leq 20$
Category V: $20 < \varepsilon \leq 25$
Category VI: $\varepsilon > 25$

Table 1.1 Scores for groundwater table depth H.

Depth range	$H \leq 10\,m$	$10\,m < H \leq 20\,m$	$20\,m < H \leq 30\,m$	$30\,m < H \leq 40\,m$	$H > 40\,m$
Score	1	2	3	4	5

Table 1.2 Scores for low-permeable deposit thickness and lithology.

	Lithology Group of Deposits/(Hydraulic Conductivity, k, m/day)		
Thickness of Low-Permeable Deposits, m_0, m	A $(0.01 \leq k < 0.1)$, Loamy Sands, Light Sandy Loams	B $(0.001 \leq k < 0.01)$, Mixed A and C	C $(k < 0.001)$, Heavy Sandy Loams, Clays
$m_0 \leq 2$	1	1	2
$2 < m_0 \leq 4$	2	3	4
$4 < m_0 \leq 6$	3	4	6
$6 < m_0 \leq 8$	4	6	8
$8 < m_0 \leq 10$	5	7	10
$10 < m_0 \leq 12$	6	9	12
$12 < m_0 \leq 14$	7	10	14
$14 < m_0 \leq 16$	8	12	16
$16 < m_0 \leq 18$	9	13	18
$18 < m_0 \leq 20$	10	15	20
$m_0 > 20$	12	18	25

   Less favorable groundwater protectability conditions correspond to category I, and most favorable ones correspond to category VI. Suppose, for example, that the groundwater table is at depth 7 m from the land surface (score 1 according to Table 1) and there is a 3 m thick layer of loamy sand and light sandy loam in the unsaturated zone (score 2 by lithology group A, Table 1.2). Then, by the sum of the scores $\varepsilon = 3$, the groundwater protectability category is *I*. If the groundwater table is at depth 14 m (score 2, Table 1.1) and there is a 5 m thick layer of clays (score 6 by group C, Table 1.2), then $\varepsilon = 8$, which corresponds to the groundwater protectability category II.

   Confined groundwater protectability can be assessed using Goldberg's qualitative groundwater protectability assessment by the thickness of the overlying (low-permeable) confining bed, m_0, also taking into account the data on the ratio of groundwater hydraulic head in the confined and upper unconfined aquifers. If the hydraulic conductivity k_0 of the confining bed is known, then a more refined groundwater protectability assessment can be performed using the parameter $\alpha = m_0/k_0$, physically determining the water percolation time through the confining bed at a unit groundwater hydraulic head gradient (flow directed downward). Taking k_0 as ranging from

10^{-5} to 10^{-3} m/day and characteristic m_0 values of 5, 10, 20, 30, 40, and 50 m, Goldberg obtained the range of $\alpha = m_0/k_0$ to be approximately $10^3 - 10^7$ days, and determined six categories for the confined groundwater protectability assessment as follows:

Category I: $m_0 \leq 5$ m, or $\alpha \leq 10^3$

Category II: 5 m $< m_0 \leq 10$ m, or $10^3 < \alpha \leq 10^4$

Category III: 10 m $< m_0 \leq 20$ m, or $10^4 < \alpha \leq 10^5$

Category IV: 20 m $< m_0 \leq 30$ m, or $10^5 < \alpha \leq 10^6$

Category V: 30 m $< m_0 \leq 50$ m, or $10^6 < \alpha \leq 10^7$

Category VI: $m_0 > 50$ m, or $\alpha > 10^7$

The higher the category, the higher the groundwater protectability.

In addition, Goldberg determined three basic groups of confined groundwater protectability based on the confining bed thickness m_0 and ratio of groundwater heads (levels) in the upper (unconfined) aquifer, H_1, and in the assessed confined aquifer, H_2:

I. *Protected.* The groundwater is confined by a continuous (in area) permeability formation with thickness $m_0 > 10$ m and $H_2 > H_1$.

II. *Conditionally Protected.* The groundwater is confined by a continuous (in area) low-permeability formation with thickness 5 m $\leq m_0 < 10$ m and $H_2 > H_1$ (case *a*) or thickness $m_0 > 10$ m and $H_2 \leq H_1$ (case *b*).

III. *Unprotected.* The groundwater is confined by a thin confining formation with $m_0 < 5$ m and $H_2 \leq H_1$ (case *a*) or when the confining formation is discontinuities (presence of lithological "windows," zones of intensive fracturing, faults) at any ratio between H_2 and H_1 (case *b*).

Confined groundwater should also be considered as *unprotected* in the following cases: in the river valleys when the confining layer is cut through by the river in the karst areas when the confining layer is subjected to karst processes, and under the unfavorable tectonic conditions (presence of intensive neotectonic movements in the active water exchange zone, high conductivities in faults).

In group I the groundwater protectability is guaranteed by the high thickness of the confining layer and by hydrodynamic conditions at which the downward groundwater flow from the unconfined aquifer is impossible.

A quantitative *upper groundwater protectability assessment* by Goldberg is performed directly by the calculation of percolation time t_w using formula (1.1), (1.2), or (1.3). Setting the base at the *maximum contaminant lifetime*, which is assessed to be 400 days for most of bacteria, and some kinds of pesticide contamination, Goldberg determined six groundwater protectability categories, as given in Table 1.3.

For the *confined groundwater*, the time of groundwater percolation through the confining bed (at downward flow direction, $H_1 > H_2$) is calculated as

$$t_w = \frac{m_0^2 n}{k_0 \, \Delta H}, \tag{1.6}$$

Table 1.3 Categories of Goldberg's quantitative upper groundwater protectability assessment.

Groundwater protectability category	I	II	III	IV	V	VI
Percolation time t_w, days	$t_w \leq 10$	$10 < t_w \leq 50$	$50 < t_w \leq 100$	$100 < t_w \leq 200$	$200 < t_w \leq 400$	$t_w > 400$

Table 1.4 Groups and gradations of confined groundwater protectability by percolation time t_w through low-permeable confining bed.

Groundwater protectability group	Unprotected	Conventionally protected			
Gradations t_w, years	1 $t_w < 1$	2 $1 < t_w \leq 5$	3 $5 < t_w \leq 10$	4 $10 < t_w \leq 20$	5 $t_w > 20$

where n is the porosity of the confining bed (usually taken to be 0.01). Corresponding confined groundwater protectability gradations and groups are given in Table 1.4.

Comparing the assessed t_w value with known lifetimes for definite contaminants, a special groundwater protectability assessment can be done for these contaminants.

The groundwater protectability assessment system described above and developed by Goldberg was the first theoretically grounded solution of the given problem. The system has been used with different modifications and generalizations in Russia until recent times [*Goman*, 2005; *Michnevich*, 2011].

It should be noted, however, that a groundwater protectability assessment based on the time of water percolation through the overlying deposits determined for different cases by formulas (1.2)–(1.4) and (1.6) is in most cases not complete enough because it assesses only *cover* groundwater protectability and does not account for the protective capacity of the aquifer itself.

In the case of infiltration of contaminated water from the surface, calculations of the water percolation time t_w through the unsaturated zone by formula (1.3) show that the assessed percolation time appears to be small enough. For example, at infiltration rate $w_c = 100$ mm/year and a 10 m thick unsaturated zone with effective porosity 0.01 composed of heavy loams and clays with hydraulic conductivity 0.001 m/day, formula (1.3) gives a t_w equal to only 239 days. This result is in agreement with the conclusion by *Haustov* [2007] that the cover protectability of upper groundwater, even in a thick unsaturated zone (over 10 m) composed of low-permeability deposits (loams, clays) is always insufficient for groundwater protection from contaminants. For this reason, the upper groundwater can never be "well protected" or "protected enough" but only "relatively" or "conditionally."

Thus, although the groundwater protectability assessment by water percolation time from the land surface to the groundwater table accounts for the hydraulic conductivity of covering deposits, this method does not account for the presence of geochemical barriers as well as the hydraulic and geochemical capacity properties of the assessed aquifer itself.

After Goldberg, the development of groundwater protectability assessment methods in the former USSR is associated with the works of *Mironenko and Rumynin* [1990, 1999], *Pashkovskiy* [2002], *Pityeva* [1999], and *Zektser* [2001]. Their efforts were directed at accounting not only for hydraulic properties but also for physicochemical properties of soil, in both unsaturated and saturated zones.

In particular, *Mironenko and Rumynin* [1990] determined the percolation time t_w of a conservative contaminant from the soil surface to the groundwater by the balance equation

$$wt_w = \int_0^{m_A} \theta(z)\,dz, \tag{1.7}$$

where $\theta(z)$ is the volumetric water content that can in turn be related to infiltration w, full moisture saturation θ_m (equal to effective porosity), field capacity θ_0 of soil (water content held in soil after excess water has drained) and hydraulic conductivity k by the formula

$$\theta = \theta_0 + \left(\theta_m - \theta_0\right)\sqrt[4]{\frac{w}{k}}. \tag{1.8}$$

Rumynin [2003] studied the sorption properties of groundwater geological medium and their effect on radionuclide migration.

Pashkovskiy [2002] proposed to assess the *contaminant travel time* t_c taking into account sorption in soil and the unsaturated zone:

$$t_c = \frac{mR}{w}, \quad R = 1 + K_d \frac{\delta}{\vartheta}, \tag{1.9}$$

where m (m) is the thickness of the unsaturated zone, K_d (dm³/kg) the distribution coefficient, δ (kg/dm³) the volume (specific) weight of rock, θ the volumetric water content (usually substituted by effective porosity n), w (m/day) the infiltration velocity, and R the retardation factor determined as the ratio of water and contaminant velocities.

Zektser [2001] used the same approach to determine the contaminant travel time in the unsaturated zone. The author also introduced the concepts of the *full residence time* of a contaminant and the *time of water exchange* in the groundwater system considered based on the balance of the groundwater recharge and discharge. Considering a more general approach to the groundwater protectability assessment, Zektser also gave a more generalized determination of groundwater protectability as the property of a natural system which allows the groundwater composition and quality to be preserved as satisfying the requirements of the groundwater practical use during a forecast period. This means that requirements for groundwater protectability are different depending on its use, e.g., for potable, technical, or industrial purposes.

For a groundwater protectability assessment in any groundwater system (saturated or unsaturated), *Zektser* [2001] *and Rogachevskaya* [2002] determined the full residence times T_w and T_c for nonsorbed and sorbed contaminants, respectively, by the formulas

$$T_w = V/Q, \tag{1.10}$$

$$T_c = VR/Q, \tag{1.11}$$

where V is the volume of the system, Q is the rate of groundwater flow passing through the system, and R is the retardation factor determined by equation (1.9).

The geochemical aspects of groundwater were studied *Kraynov and Shvets* [1987], *Kraynov et al.* [2004], *Pityeva* [1999], and *Pityeva et al.* [2006] based on the concept of geochemical barriers of geological medium. This concept was first proposed by *Perelman* [1961], who determined the geochemical barrier as a zone in which a sharp change of hydrogeochemical conditions of chemical element migration takes place at short distances, causing their precipitation to a solid phase.

Pityeva [1999] proposed the concept of "geochemical groundwater protectability" determined by a series of physicochemical processes causing the removal of contaminants from the groundwater, such as sorption in porous or fractured media. According to Pityeva, geochemical groundwater protectability includes:
- identification and quantitative assessment of physicochemical processes along the travel paths of contaminants to groundwater;
- their;
- assessment of the potential manifestation of these processes in different conditions and objects determining the groundwater protectability.

Assessment of groundwater protectability is conducted the depending on types and properties of water-bearing rocks, as well as the thickness of the unsaturated zone.

Further development of the hydrogeochemical aspects of groundwater vulnerability assessment is found in the work by *Goman* [2007] as related to the migration of organic contaminants through low-permeable hydrogeological beds in areas of common solid waste repositories.

"The *Russian methodology"* [*Belousova and Galaktionova*, 1994; *Belousova*, 2005]. The Chernobyl catastrophe provided significant amount of information on a large scale on groundwater contamination with radionuclides, in particular with ^{137}Cs and ^{90}Sr [*Shestopalov*, 2001, 2002]. The accident groundwater protectability assessments have been implemented for Chernobyl-born ^{137}Cs by *Belousova and Galaktionova* [1994] based on the contaminant travel time through the unsaturated zone taking into account its thickness, lithology, and sorption properties. The same approach was used for a regional assessment of upper groundwater vulnerability to Chernobyl-born ^{137}Cs for the Dnieper Basin areas of Ukraine and Russia performed during the Russian-Belorussian-Ukrainian Cooperated Research Project in 2003 [*Shestopalov*, 2003]. As part of this research we used the base Russian methodology of intrinsic groundwater vulnerability assessment [*Belousova and Galaktionova*, 1994]. As a result, a regional groundwater vulnerability assessment by contamination was performed for the area of the Kyiv region, including the Chernobyl Exclusion Zone (CEZ), and a groundwater vulnerability map in scale 1:20,0000 was drawn. The methodology is based on the assessment of the contaminant travel time from the contaminated surface to the groundwater table, t_c according to formula (1.9) [*Pashkovsky*, 2002] applicable for both conservative and sorbed pollutants. Depending on the t_c value, the score scale for the upper groundwater vulnerability to ^{137}Cs is determined as shown in the Table 1.5.

As shown in the Table 1.5, groundwater vulnerability is classified into seven categories: catastrophic, very high, high, medium, low, very low, and absent. The two lowest categories, very low and absent, are often unified as "conditionally invulnerable."

Rogachevskaya [2002] used materials obtained in the above study as well as data from field observation of ^{137}Cs migration in the unsaturated zone obtained at special test sites. She considered groundwater vulnerability as a concept inverse to groundwater protectability based on the hydraulic and geochemical barriers of the unsaturated zone, influence of forestation degree as a regional factor, and hydrogeological properties of the assessed upper aquifer. As important factors of groundwater vulnerability to radioactive contamination, the sorption capacity

Table 1.5 Gradations of groundwater vulnerability by ^{137}Cs as determined by surface contamination density (Ci/km^2) and radionuclide travel time t_c from surface to groundwater table.

t_c Range, years	Groundwater Vulnerability Grade				
	>40 Ci/km^2	15–40 Ci/km^2	5–15 Ci/km^2	1–5 Ci/km^2	<1 Ci/km^2
$t_c < 30$	Catastrophic	Very high	High	Medium	Very low
$30 < t_c < 60$	Very high	Very high	High	Medium	Very low
$60 < t_c < 100$	Very high	High	Medium	Low	Very low
$t_c > 100$	Medium	Low	Low	Low	Absent

(retardation), dispersion, and radioactive decay are considered. As the basic parameter of groundwater vulnerability assessment, *Rogachevskaya* [2002] used the radionuclide full residence time T_c in the hydrogeological system as determined above by equations (1.10) and (1.11):

$$T_c = T_w R, \qquad (1.12)$$

where, as before, T_w is the residence time in the hydrogeological system of a nonsorbed chemicals moving with groundwater flow velocity, and R is the retardation factor determined by equation (1.9). During construction of the resulting groundwater vulnerability map, the zoning map of protective properties for the unsaturated zone is overlaid with the map of radionuclide residence time in groundwater determining the self-cleaning aquifer ability.

Based on results of experimental studies of ^{137}Cs migration in areas contaminated after the Chernobyl Nuclear Power Plant (NPP) accident in Russia (Bryansk region) and experiments with artificial radionuclide injection at special observation plots, Rogachevskaya came to the conclusion that the soil is not a perfect protective barrier against radionuclide migration from the soil surface to groundwater. The share of "fast migration component" of the nonsorbed contaminant appeared to be near 10%. This part is determined by fast migration pathways such as "breakthrough" pores of the unsaturated zone and local "migration windows." Of key importance are the relief microforms which influence the infiltration and depot properties of the soil and unsaturated zone.

The above conclusions are in agreement with our representation of the existence and importance of preferential flow and migration zones (PFMZs) of different scales in the geological medium. The assessed share of PFMZs in the total groundwater contamination (10% from total initial contamination) determined just on the local site scale (without accounting for larger PFMZs such as depressions and lineaments) is rather significant. Moreover, the effects of the landscape type (forested, meadow, plowed, etc.) also provide important input into the assessment of groundwater protectability.

Polyakov and Golubkova [2007] also used the water exchange time and retardation factor. However, they estimated the residence time of a nonsorbed tracer (or water exchange time), T_w using a nonsorbed radioactive tracer (tritium). The tritium concentration was measured, and the time T_w was determined by "input" and observed tritium concentrations according to the methodology proposed by *Maloszevski and Zuber* [1996]. As an "input function," they used historical data on tritium concentration in atmospheric precipitation starting from 1953 (when nuclear weapon tests were conducted in the atmosphere). The authors accounted for the retardation factor and lifetime of the radionuclide. They developed a score assessment system for groundwater vulnerability as applied to the area of Azov-Kuban artesian basin (score range 0–7) corresponding to the average water exchange time from over 1000 years to 5 years and determined tritium concentrations from

1 to 14 TU (tritium units, 1 TU $= 0.119$ Bq/L). The wide use of this method requires implementation of special field works for groundwater sampling and sample analysis for determination of isotopes Tr, δD, δ^{18}O, δ^{13}C, and δ^{14}C.

AVI method. Among the parametric groundwater vulnerability assessment methods, one should mention the aquifer vulnerability index (AVI), which was developed at the National Hydrogeology Research Institute of Saskatoon (Canada) by *Van Stempvoort et al.* [1995]. The authors used the total flow resistance of the covering deposits taking into account the lithology:

$$r = \sum_i \frac{m_i}{k_i}, \tag{1.13}$$

where m_i are layer thicknesses and k_i are the corresponding hydraulic conductivities. The method is equivalent to the assessment using groundwater percolation time because the total resistance r can be treated as the time of water percolation through the whole formation at a unit vertical hydraulic head gradient.

The method was by *Tovar and Rodriguez* [2004] for a groundwater vulnerability assessment in the area of Leon, Mexico. The hydraulic conductivities were determined by pumping tests and direct measurements with a constant head permeameter. The assessment required a significant volume of initial information that was provided by detailed GIS maps of relief, geological conditions, and conditions of land use. The obtained results were compared with an alternative assessment using the DRASTIC method. Authors noted that the AVI method gave a higher vulnerability, particularly in zones of tectonic dislocations.

Overall, parametric groundwater vulnerability assessments by water percolation time or flow often lead to underestimation or overestimation of the potential groundwater contamination depending on whether or not the physicochemical interaction in the "contaminant-water-rock" system is taken into account. The approach uses the representation of a contamination front with a definite concentration at a definite depth below which the groundwater medium is still considered clean at each time moment. The approach often takes no account of areal distribution PFMZs of different dimensions (from macropores to areal zones related with depressions, geodynamically active zones, etc.). With an increase in the assessed area, the heterogeneities of larger dimensions should be brought into consideration, and the assessed groundwater vulnerability should be determined by their total "degree of openness."

1.4. Modeling Methods

Among the modeling methods used by different authors for groundwater vulnerability assessments, two methods should be distinguished: deterministic and statistical.

Most deterministic methods are based on general flow and transport balance (conservation) equations for the modeling domain with corresponding boundary conditions determining a boundary or initial-boundary (in the transient case) problem, using a system of partial differential (or integral) equations. These boundary problems mathematically describe the principal physical processes determining contaminant transport in a water-bearing system, the most important of which are *advection* (transport with groundwater flow velocity), *dispersion* of the contaminant front (caused by different deviations of contaminant particles from their "advection" paths and positions), and *sorption* of the contaminant by water-bearing rock.

The partial differential equation describing the contaminant transport in groundwater in saturated conditions can be written in the form [*Ciang and Kinzelbach*, 2001]

$$\frac{\partial C}{\partial t} = \frac{\partial}{\partial x_i}\left(D_{ij}\frac{\partial C}{\partial x_j}\right) - \frac{\partial}{\partial x_i}(v_i C) + \frac{q_s}{n}C_s + \sum_{k=1}^{N}R_k, \qquad (1.14)$$

where C is the concentration of a dissolved contaminant in groundwater (in units of mass or activity per unit volume, M/L^3), t is time (T), x_i are linear distances along the corresponding axes of the Cartesian coordinate system (L), D_{ij} is the hydrodynamic dispersion tensor ($L^2 T^{-1}$), v_i is the real flow velocity (LT^{-1}), q_s is the volume water flow rate per unit volume of water-bearing medium representing sources of water recharge and discharge (T^{-1}), C_s is the contaminant concentration in the recharge and discharge sources (ML^{-3}), n is the porosity (dimensionless), and $\sum_{k=1}^{N}R_k$ is a chemical reaction term, or the contaminant mass recharge or discharge sources ($ML^{-3}T^{-1}$).

When only the equilibrium sorption and irreversible reactions of first-order chemical reactions are considered, the chemical reaction term in equation (1.14) can be represented in the form [*Grove and Stollenwerk*, 1984]

$$\sum_{k=1}^{N}R_k = -\frac{\rho_b}{n}\frac{\partial \overline{C}}{\partial t} - \lambda\left(C + \frac{\rho_b}{n}\overline{C}\right), \qquad (1.15)$$

where ρ_b is the specific weight of rock (mL^{-3}), $\overline{C}$ the concentration of contaminant sorbed by rock per unit rock mass (Mm^{-1}), and λ the first-order chemical reaction rate constant (T^{-1}). The contaminant transport equation (1.14) is coupled with the groundwater flow equation by the relation

$$v_i = -\frac{K_{ii}}{n}\frac{\partial h}{\partial x_i}, \qquad (1.16)$$

where K_{ii} is the main component of the hydraulic conductivity tensor (L/T) and h is the hydraulic head (L). The hydraulic head distribution is determined by the groundwater flow equation

$$\frac{\partial}{\partial x_i}\left(K_{ii} \frac{\partial h}{\partial x_j} \right) + q_s = S_s \frac{\partial h}{\partial t}, \tag{1.17}$$

where S_s (L^{-1}) is the specific storage coefficient (storativity, or specific yield) of a water-bearing porous medium.

For numeric solution of the 3D boundary problems of groundwater flow and transport described by equations (1.14)–(1.17), various computer codes have been developed. The most well-known are the MODFLOW code for groundwater flow [*McDonald and Harbaugh*, 1988] and the MT3D code for contaminant transport [*Zheng*, 1990].

Depending on the case assessment and its objectives, different simplified versions of the 3D equation system (1.14)–(1.17) can be used: 1D (vertical), 2D (cross-section), etc.

For example, *Zhang et al.* [1996] assessed the intrinsic groundwater vulnerability of the Goshen County, Wyoming, using a 1D advection-dispersion model for the unsaturated zone. To determine the vertical distribution of the contaminant concentration, they solved a 1D equation of the type 1.14 without accounting for sorption in which, instead of porosity, they considered the water content as a function of the hydraulic conductivity of *van Genuchten* [1980]. The governing equations of water flow and chemical transport with the specified initial and boundary conditions were solved using a computer code HYDRUS (developed at the US Salinity Laboratory) using the finite-element method [*Vogel et al.*, 1995]. The authors calculated 130 vertical concentration distributions of the relative contaminant concentration c/c_0, where c_0 is contaminant concentration in water infiltrating from the surface. The resulting groundwater vulnerability assessment has been compared to the corresponding assessment using the modified DRASTIC method (with procedures of GIS map overlays). The authors [*Vogel et al.*, 1995] note that the index-rating methods with GIS are appropriate for large study areas and the modeling method is better for use at smaller sites.

In the work of *Loague et al.* [1998] a 3D model is developed based on the MODFLOW-MT3D code for the regional groundwater vulnerability assessment of Fresno County, California, to contamination with DBCP (1,2-dibromo-3-chloropropane), which was used since 1940 until its prohibition in 1977. The authors reconstructed the historic data on atmospheric precipitation, land use (state of soil), irrigation, and groundwater contamination with DBCP and developed a 3D groundwater flow and contaminant transport model. In the result of the epignostic (past-time) simulation for the period 1960–1994 the authors built the map of groundwater contamination with the pesticide. The results obtained in this work were further analyzed by *Loague and Corwin* [1998], they came to the conclusion that 3D modeling using GIS technologies is, in many cases, most effective for groundwater vulnerability assessment. The GIS provides the direct data support for modeling (preprocessing, postprocessing, reformatting, mapping, etc.), especially in the analysis of non–point source vulnerability. It helps to characterize the full information content of the spatially variable data required by solute transport models.

The same conclusion is made by *Zaporozec* [1985] as the result of a groundwater vulnerability assessment in Wisconsin using the SUPRA index-rating method. He notes that the step after the preliminary assessment should be development of a regional hydrogeological flow transport model for the study area.

Statistical models of groundwater flow and transport used for groundwater vulnerability assessments are in most cases equivalent to deterministic ones because their general solutions also satisfy the groundwater flow and contaminant mass balance equations. However, the problem solution methods are based on stochastic algorithms such as the Monte Carlo method. Another aspect of statistical models is represented by the use of special probability density functions for the solution of groundwater migration problems. For groundwater modeling applications, this approach was developed by *Jury and Roth* [1990].

Statistical algorithms and data processing methods (regression analysis, interpolation and extrapolation methods, gridding methods, etc.) are directly employed in groundwater vulnerability assessment *by analogy*, that is, by associating a given research area with known areas in which groundwater contamination already occurred. If the analogue area is the same as the studied area, then we have the case of groundwater vulnerability assessment *by real contamination*. For example, *Evans and Maidment* [1995] used such a method for a statistical assessment of the groundwater vulnerability in Texas to nitrate contamination using linear regression analysis. They built a spatial distribution map for groundwater contamination probability based on water sampling data from 29,485 wells in the study area. It is clear that this method requires high volumes of initial information available only using the monitoring network facilities. An extended review of statistical methods for groundwater vulnerability assessment is presented in the National Research Council reports [*NRC*, 1993a,b].

Among all groundwater vulnerability assessment methods described above, in a higher or lesser degree, only a few consider the pathways and zones of preferential flow and transport. An attempt at the experimental assessment of a "fast migration component" for Chernobyl-born ^{137}Cs was implemented by *Rogachevskaya* [2002]. In other methodologies the preferential flow phenomena were taken into account indirectly, particularly in the German, EPIC, PI, and COP methods for karst areas [*Hoelting et al.*, 1995; *Doerfliger et al.*, 1999; *Zwahlen*, 2004].

In the present work an attempt is made to assess the groundwater vulnerability of the upper (Quaternary) and first confined (Eocene) aquifers of the Dnieper River basin area (Kyiv region) in Ukraine to contamination with Chernobyl-born ^{137}Cs, taking into account the PFMZ associated with depressions. A methodology is proposed for this purpose based on the 1D contaminant transport model and 3D groundwater flow model of the study area as well as available data of observed groundwater contamination obtained during the postaccident period.

As *Shestakov* [2003] notes, it is necessary to formulate the question of possible manifestations of heterogeneity in the geological medium during the solution of any hydrogeodynamic problem. Any study of hydrogeodynamic processes cannot be considered complete if the influences of rock heterogeneity on these processes have not been analyzed. This conclusion is especially acute in questions of groundwater vulnerability and protectability assessments.

2. CHERNOBYL-BORN RADIONUCLIDES IN GEOLOGICAL ENVIRONMENT

The questions of radionuclide migration assessment in the geological medium (groundwater and water-bearing rocks) after the Chernobyl accident are complicated, from one side, because of essentially inhomogeneous degree of contamination of landscapes and, from the other side, the high complexity and variability of the structure of the geological environment within the studied region, which are complicated by different human-induced factors influencing the migration processes in the active water exchange zone.

One of the main objectives of groundwater vulnerability and protectability assessment is obtaining a preliminary forecast of vertical penetration of radioactive contaminants into the multi-layered aquifer system through the unsaturated zone.

The preferential downward radionuclide transport within the CEZ and Kyiv region is explained by several reasons, primarily the following:

1. Presence of low relief zones within the contamination source area, that is, the CEZ with rather small surface runoff into the river network (river Pripyat and its tributaries)
2. Good vertical permeability of the unsaturated zone because of preferentially sandy rocks with presence of turf insertions
3. Presence of active vertical preferential flow pathways caused by specific conditions within numerous closed relief depressions accumulating local surface runoff and contamination
4. Existing climatic conditions characterized by intensive precipitation and moderate evaporation

The role of the vertical flow component remains to be primary also for prediction assessments of contamination of deeper aquifers. This conclusion is obvious because of relatively small lateral flow velocities as compared to vertical ones if taken relative to corresponding dimensions of the aquifer system. Nevertheless, for assessments of the contamination of deeper confined aquifers in the presence of vertical preferential flow pathways, the determination of

Groundwater Vulnerability: Chernobyl Nuclear Disaster, Monograph Number 69.
Edited by Boris Faybishenko and Thomas Nicholson.
© 2015 American Geophysical Union. Published 2015 by John Wiley & Sons, Inc.

groundwater flow parameters and solution of the whole (plane and vertical) problem is important. In particular, this is explained by the necessity of regional long-term predictions of groundwater resources and quality, especially in conditions of their intensive exploitation during the postaccidental period.

Starting from the concept of the existence of downward radionuclide transport from the surface with infiltration water, the geological rocks along the filtration pathway could accumulate definite amounts of radionuclides, in such a way being a protection barrier as well as a potential secondary source of contamination for deeper aquifers.

After the Chernobyl accident, the surface contamination with ^{137}Cs and ^{90}Sr in the vicinity of the Chernobyl NPP reached several MBq per square meter and several tens of kBq at farther distances.

According to the results of studies implemented soon after the Chernobyl NPP accident in April 1986 [*Borzilov*, 1989; *Baryakhtar et al.*, 1997], it was assumed that radioactive contamination coming to the soil surface by atmospheric fallout is concentrated in the upper soil layer, and its lower boundary gradually deepens with time. It is obvious that different-scale heterogeneities of the geological medium as well as relief variability result in the existence of various preferential pathways of water infiltration and radionuclide migration from the contaminated soil into the groundwater. Such pathways are distinguished from the average "background" geological medium by the intensity of the migration process as determined by different velocities of downward radionuclide migration from the upper soil layers into the unsaturated zone and groundwater aquifers.

The separate sampling results of radionuclide concentrations in the soil layer and unsaturated zone show that in the most studied areas these concentrations are relatively small and usually do not exceed 1% of the initial surface concentration. For this reason, they are not taken into account in the balance calculations. However, when accounting for a possible occurrence of PFMZs, the total transported amount of radionuclides into groundwater can become significant even at a relatively high depth.

This preliminary conclusion has been confirmed by the results of postaccident studies of groundwater contamination with Chernobyl-related radionuclides ^{90}Sr and ^{137}Cs [*Shestopalov et al.*, 1992, 1996, 1997].

The results of field research and experiments implemented in the study area of the Kyiv region and CEZ showed that the most probable PFMZ occurrence is related to closed depressions in the relief, which exert a significant influence on the intensity of vertical infiltration and transport of radionuclides from the contaminated surface into groundwater. Their occurrence and activity increase in the study area from south to north and northeast in relation with changing the types of geomorphological conditions.

The southern area (to the latitude of Kyiv on the right bank of the Dnieper) is located in the Dnieper Upland. Its unsaturated zone is composed of black soils and loams. Depression landforms are absent or very rare. Linear forms dominate, such as gullies and ravines, especially deep in the side part of the Dnieper Upland.

Central areas on the right bank of the Dnieper have reduced smoothed terrain, loamy sod-podzolic soils, and sandy-loamy composition of the aeration zone. Depressions forms are available here.

Northern and eastern regions of the left bank of the Dnieper are related to the sandur outwash plains of Ukrainian Polesye and first-second terraces of the Dnieper and Pripyat [*Shestopalov*, 2001].

Characteristic for the sandur plains are water-glacial, aeolian, and partially glacial deposits with sandy, loamy-sandy, and sandy-loamy composition. Their surface is mainly represented by aeolian positive (swells, elongated sandy ranges, cirques with relative elevation 5–12 m) and negative (sinks, interrange cauldrons, coombs) relief forms. Less characteristic are ancient forms of relief such as humps, hills, and depressions of preglacial and glacial origin. With increasing distance from the Pripyat and Uzh rivers toward the watershed, the surface of sandur plains becomes more flat. The absolute elevation of sandur plains ranges from 120 to 140 m, and the average density of erosion network is about 0.2 km/km^2. Because of the low-developed erosion (drainage) network, light mechanical composition of soil cover, and large number of closed low-runoff areas, the area is characterized by low surface runoff which slightly increases from west to east.

Terraces of the Dnieper and Pripyat are also plains. Their alluvial deposits consist of interbedded sands with different grain size and sandy loams. The depressions are widely spread here, especially in the eastern part of the territory on the terraces of the Dnieper.

The hydrogeological conditions of the study area (Kyiv region and the CEZ) are characterized by the presence of four principal aquifers (water-bearing complexes): (1) Quaternary (depth to 30 m), (2) Eocene (depth to 100 m), (3) Cenomanian-Callovian (depth to 150 m), and (4) Bajocean (depth to 280 m).

The Quaternary aquifer is related to the recent alluvial deposits of floodplains and river beds, alluvial and lacustrine deposits of the first and second floodplain terraces, and fluvioglacial, lacustrine-glacial, and lacustrine-alluvial deposits. The deposits are represented by quartz sands of different grain size often interlaid with loamy sands, sandy loams, and clays. This aquifer is of general occurrence over the studied area. It is underlain by a low-permeable layer of Neogene red clay and in the places of its absence by water-bearing sandy Oligocene-Pliocene deposits.

The Eocene water-bearing complex is of general occurrence over the area. The deposits are composed of quartz sands. At the base of the Eocene aquifer, the Upper Cretaceous marl-chalk layer is present, being the regional low-permeable bed. Almost over the whole territory the Eocene aquifer is overlain by a low-permeable layer of Kyiv suite composed of marl and siltstone. The water head above the top of the aquifer ranges from 8 to 50–80 m. In the late 1960s a depression cone in the groundwater levels was formed around Kyiv City caused by water intake wells exploiting the underlying aquifers for water supply to Kyiv. The aquifer is utilized for water supply to cattle farms in the rural area and satellite settlements of Kyiv.

The water-bearing complex in Cenomanian (Lower Cretaceous) and Callovian (Middle Jurassic) deposits occurs at depths from 80 to 150m over the whole study area excluding the southwestern part of the Kyiv urban agglomeration. The water-bearing rocks are represented by sands of different granularity sometimes interlaid with sandstone and limestone. The aquifer is commonly overlain by the low-permeable layer of the Upper Cretaceous marls and chalks being absent only in the southern part of the studied area. At its base, a low-permeable layer of Middle Jurassic clays and siltstone is present. The aquifer is one of the principal sources for potable water supply to Kyiv City. At present the total pumping rate reaches 200,000 m³/day.

The Bajocean (Middle Jurassic) aquifer is also of regional occurrence within the study area. The water-bearing rocks are represented by sands. The thickness of sandy deposits rises in the eastern direction, ranging from a few meters to 40–60 m. The aquifer is confined and the head can reach 280m. It is one of two main aquifers being exploited for water supply to Kyiv City. As a result of exploitation, a depression cone of 60 km radius was formed in the aquifer with the head drawdown reaching 120m in the center.

Depending on hydrogeological and man-induced conditions (e.g., water intake operation), the Chernobyl-related groundwater contamination by ^{137}Cs and ^{90}Sr was observed at different concentrations in practically all aquifers in the regions where the surface contamination initially occurred. According to sampling data obtained in 1992–1997 (over 700 ^{137}Cs and 500 ^{90}Sr samples taken from wells in the Kyiv and CEZ regions except for the immediate vicinity of the Chernobyl NPP), the concentrations in the groundwater of the main aquifers used for the water supply reached 100 mBq/dm³ and more (Table 2.1).

Table 2.1 Concentrations of ^{137}Cs and ^{90}Sr in the groundwater of Kyiv regional aquifers given as the percentage of total samples in the aquifer for given concentration ranges. Data collected in 1992–1997.

Aquifer (Age)	Sampling Depth Interval, m	^{137}Cs				^{90}Sr		
		<10 mBq/dm³	10–50 mBq/dm³	51–150 mBq/dm³	>150 mBq/dm³	<10 mBq/dm³	10–50 mBq/dm³	>50 mBq/dm³
Quaternary	2–30	41%	44%	11%	4%	53%	42%	5%
Eocene	45–65	43%	45%	9%	3%	74%	21%	5%
Cenomanian-Callovian	80–150	50%	36%	8%	6%	80%	20%	–
Bajocean	200–280	51%	36%	7%	6%	75%	24%	1%

The areal distribution of ^{137}Cs and ^{90}Sr concentrations in the groundwater of the upper Quaternary aquifer (depths to 30 m) within the Kyiv region as plotted for 1996 is shown in Figure 2.1.

In the Quaternary aquifer, the maximum groundwater contamination with ^{137}Cs is observed in the northern part of the region, showing a good correlation with the surface contamination density (Figure 2.2), which increases nearing the Chernobyl NPP.

For ^{90}Sr, which is characterized by its higher migration ability as compared to ^{137}Cs, relatively high concentrations were found in groundwater not only in the northern part of the region but also around Kyiv City, where the depression cone generated by intensive exploitation of the groundwater of deeper (Cenomanian-Callovian and Bajocian) aquifers has led to an increased recharge intensity of the Quaternary aquifer and related downward migration of radionuclides. A groundwater sampling check in the relatively clean area with surface contamination not exceeding 20 kBq/m^2 and similar hydrogeological conditions did not reveal significant concentrations of the radionuclides.

At the end of the 1980s, measurable concentrations of the short-lived isotope ^{134}Cs (half-life two years) were discovered in the valley of the Desna River in groundwater of several artesian wells springing from the Eocene aquifer (depths 45–65 m) in the close vicinity of Kyiv. In this case it is clear that downward penetration of the radionuclide along the well casing is excluded and there is no other way of its penetration to the aquifer than vertical downward migration from the contaminated surface. This fact indicates, first, the Chernobyl-related origin of the contamination. Second, in order to penetrate to such depths, the downward velocity of the radionuclide transport should comprise about 10–15 m/year or even more. If extrapolating the possibility of downward radionuclide migration by natural pathways into the Jurassic aquifer to a depth of 250 m in conditions of the depression cone being formed due to groundwater exploitation, we obtain a migration velocity of up to 50 m/year.

Such high velocities testify to the existence of the preferential vertical flow and transport pathways in the upper sedimentary cover, probably related to disintegration zones of neotectonic activity. Lithological, mineralogical, and grain size heterogeneities of the covering deposits are also of great importance. Consequently, the discovery of radioactive isotopes and other contaminants (pesticides) of undoubtedly surface sources at relatively high depth evidences in favor of the existence of PFMZ in the upper geological environment.

In 1996–1998, a series of determinations was performed of ^{137}Cs solid-phase concentrations in core samples from specially drilled boreholes of depths down to 100 m within the Kyiv urban area. These measurements showed noticeable contamination of deposits with concentrations from 1 to 10 Bq/kg [*Shestopalov*, 2002]. The corresponding vertical concentration plotted against depth for one of these boreholes (southwest part of the Kyiv urban area) is shown in Figure 2.3.

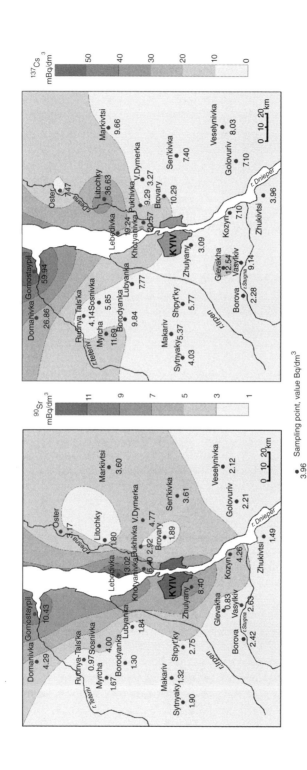

Figure 2.1 Distribution of ^{137}Cs (right) and ^{90}Sr (left) concentrations in groundwater of Quaternary aquifer in the Kyiv urban area, 1996.

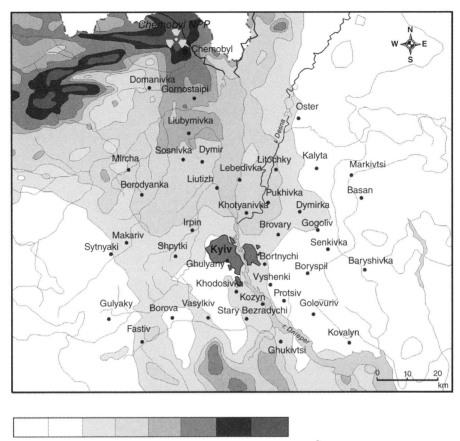

Figure 2.2 Surface contamination with ^{137}Cs in the Kyiv urban area.

A series of ^{137}Cs solid-phase concentration measurements in the core material from the Quaternary boreholes were also performed in the CEZ. A corresponding typical vertical concentration profile is shown in Figure 2.4.

It is obvious that these results cannot be considered as representative of the entire study area as they have been obtained in separate boreholes that could fall in local surface zones of high contamination density.

Along with a core sample study, several series of deposit samples were analyzed taken from the lower part of the Kyiv marl bed (depth 80 m) in the course of the Kyiv subway tunnel drilling. The results of ^{137}Cs content determinations are shown in Figure 2.5. The solid-phase concentration of the radionuclide increases along the section in the direction of the Syrets creek valley coinciding

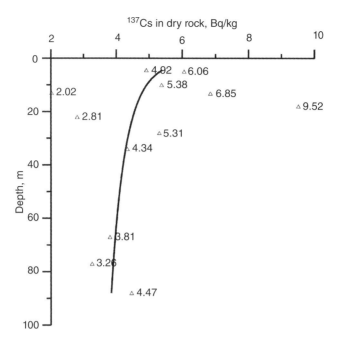

Figure 2.3 Content of ¹³⁷Cs (Bq/kg) in core samples from a borehole at the southern margin of Kyiv City.

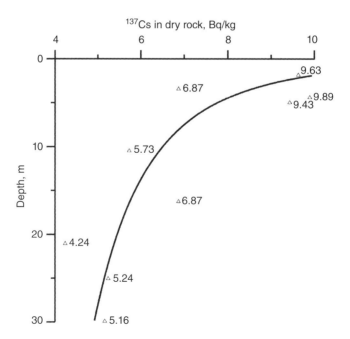

Figure 2.4 Vertical distribution of ¹³⁷Cs solid-phase concentration obtained by core samples from a borehole in the CEZ area.

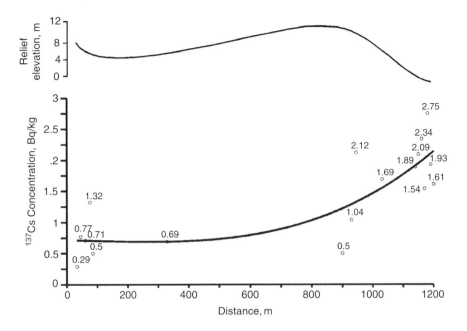

Figure 2.5 Distribution of [137]Cs (Bq/kg) in Kyiv marls (depth 80 m) plotted by sampling data obtained during drilling of tunnels for the Kyiv subway between Zoloti Vorota and Dorogozhichi stations. Upper curve shows the relief elevation.

with a linear geodynamic zone and characterized by increased permeability of the deposits. Previously an increased content of pesticides had also been discovered at this location [*Shestopalov*, 1988].

The separate available data on the real rock contamination with radionuclides and more abundant data on groundwater contamination enable to obtain a preliminary average assessment for contamination of the geological environment. With this purpose, the average distribution coefficient K_d (dm³/kg) was used, determined as the ratio of a radionuclide concentration in the solid (rock) phase M (Bq/kg) to its concentration in the liquid phase (groundwater) C (Bq/dm³):

$$K_d = M/C. \qquad (2.1)$$

The K_d values for ^{137}Cs vary in a wide range over the section from 3 to 500 dm³/kg. In close vicinities of the Chernobyl NPP it varies from several dm³/kg to several thousand dm³/kg depending on the rock properties, radionuclide migration forms (solute or colloid), groundwater chemical composition, and other factors. For this reason a reliable forecast of radionuclide concentrations in the geological environment can be obtained only after proper study of the regularities of interactions in the "soil-water-rock" system with account of its main influencing factors.

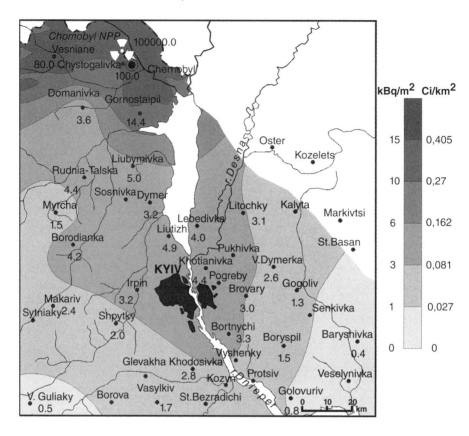

Figure 2.6 Calculated accumulation of ^{137}Cs in the geological environment of Quaternary and Eocene deposits (depth to 65 m) per unit area within the Kyiv province according to measurements performed in 2001.

However, the available information already enables researchers to obtain a preliminary calculation of ^{137}Cs accumulated per unit area (in Bq/m^2 or Ci/km^2) in the upper geological medium below the soil layer. Such a calculation was performed in 2001 for the majority of the Kyiv province area to depth 80 m. Within the CEZ area, the assessment depth varied from 25 to 60 m. The obtained distribution of accumulated ^{137}Cs activity per unit area is shown in Figure 2.6. The highest accumulation (over $15 kBq/m^2$) corresponds to the most contaminated parts of the CEZ. The area of relatively increased accumulated contamination stretches in the south direction from the Chernobyl NPP, reaching $3-6 kBq/m^2$ near Kyiv City (Figure 2.6).

The ratio of the above accumulation of ^{137}Cs in the geological medium to the initial surface contamination density plotted for the same area of the Kyiv province is shown in Figure 2.7. Close to Kyiv and its satellite settlements (Ghulyany,

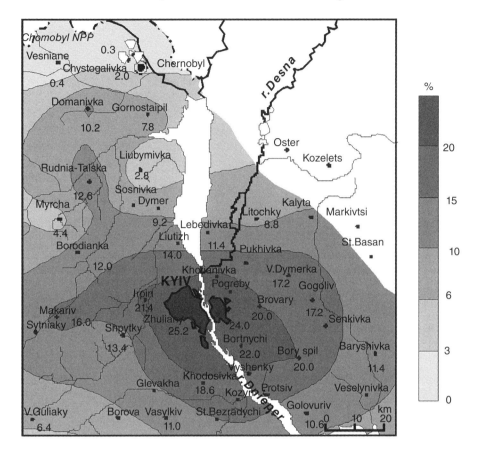

Figure 2.7 Distribution of ^{137}Cs accumulation in the geological environment (Quaternary and Eocene deposits, depth to 65 m) plotted as percentage of the initial surface contamination by data of 2001.

Bortnychi, Brovary, and Irpin) the geological environment has accumulated more than 20% of initial surface contamination. In wider areas around Kyiv (near Makariv, Borodyanka, Liutizh, Pukhivka, Baryshivka, Protsiv, Kozyn, Glevakha, and Vasylkiv) this ratio reaches 10%. Within the CEZ the ratio is below 3% because of very high surface contamination values.

An assessment of the total accumulation of ^{137}Cs in the geological environment of the CEZ has been performed for the Quaternary aquifer. Sites of the temporary locations of radioactive wastes and the Chernobyl NPP operation site with extremely high contamination have not been taken into account. According to existing data, three zones with different ^{137}Cs accumulation in the geological environment (depth to 25–40 m) per unit area have been determined, as shown in Figure 2.8.

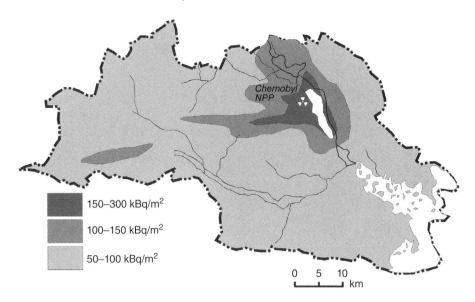

Figure 2.8 Zones of assessed ^{137}Cs accumulation recalculated per unit area for the Quaternary deposits of the CEZ by data of 2001.

In the central zone of maximum contamination (close to Chernobyl NPP), the accumulation per unit area reaches 120–300 kBq/m^2, in the intermediate zone it is about 100–150 kBq/m^2, and in the rest of the area it is 50–100 kBq/m^2.

An important factor is the annual increase of radionuclide accumulation in the geological medium. As observations show, the accumulation process proceeds with variable intensity determined by variation of atmospheric precipitation, radionuclide forms, etc. In spite of these uncertainties, the average assessment of the accumulation rate can be determined. For ^{137}Cs, at average $K_d = 20$ dm^3/kg, for the entire CEZ area it comprises about 500 Ci/year (see Table 2.2). If also considering the above-mentioned temporary locations of radioactive wastes and the Chernobyl NPP operation site, then the assessed amount of ^{137}Cs, which annually comes from the surface to the geological environment, can reach about 1000 Ci. Based on the available data, in the same way as for ^{137}Cs, an assessment of accumulation of ^{90}Sr can be obtained reaching the same order of value. As a result, the total assessed accumulation in the geological environment of ^{137}Cs and ^{90}Sr appears to be exceeding 4–23 times the annual flushing out of these radionuclides from the CEZ with water of the river Pripyat [*Kholosha et al.*, 1999].

Assuming that this process continues with the same intensity, then during 30 years (one half-life period) about 30% of the initial amount of ^{137}Cs and ^{90}Sr should be accumulated in the geological environment.

In reality, this process is rather variable over the area. Its intensity is higher within the PFMZ locations. In such locations, the rate of areal redistribution and

Table 2.2 Distribution of the total accumulation of ^{137}Cs in the geological environment of Quaternary water-bearing deposits in the CEZ.

Territory	Area (km^2)	Surface Contamination Density of ^{137}Cs		Average Total Accumulation in Upper 25 m Layer per Unit Area and Average, $K_d = 20\,dm^3/kg$			Total Area Accumulation with Correction for Radioactive Decay		Average rate of Flushing Out	
		Ci/km^2	Bq/m^2	Ci/km^2	Bq/m^2	% of surface contamination	Ci	Bq	Ci/year	Bq/year
Near NPP	20	1,000	37,000	6.2	229.0	0.006	167	6.185×10^{12}	13	4.76×10^{11}
Intermediate	200	500	18,500	4.1	152.2	0.008	1,110	3.04×10^{13}	86	3.16×10^{12}
Rest of CEZ area	1,824	100	3,700	2.5	77.2	0.02	5,071	1.39×10^{14}	390	1.44×10^{13}
Total in CEZ	2,044	—	—	—	—	—	6,348	2.35×10^{14}	489	1.809×10^{13}

transport of radionuclides from the soil surface into the deeper geological environment and groundwater significantly increases, facilitating the rehabilitation of soils and self-cleaning of areas.

From the data presented above, a conclusion can be made about the importance of studying PFMZ of different scales characterized by heterogeneities of flow and transport parameters. These zones are of principal importance in assessments of groundwater vulnerability to surface contamination as well as in questions of postaccident autorehabilitation of areas contaminated with radionuclides.

From the viewpoint of balance, physicochemical, and modeling considerations, the revelation of measurable ^{137}Cs concentrations in liquid and solid phases at significant depths is an interesting fact requiring the implementation of additional specially designed field and modeling studies. The only possible explanation for these observation results is the existence of PFMZ of different dimensions with preferential vertical pathways of downward radionuclide migration.

3. PREFERENTIAL FLOW AND MIGRATION ZONES IN GEOLOGICAL ENVIRONMENT

3.1. State of Problem Study

In spite of a large number of research works on the formation of ground-water resources [*Heath*, 1984; *Fetter*, 2000; *Vsevolzhskiy*, 1983; *Shestopalov*, 1979, 1981, 1988] and the processes of contaminant migration in the subsurface hydrosphere [*Bear*, 1972; *Fried*, 1975; *Freeze and Cherry*, 1979; *Schnoor*, 1992; *Lukner and Shestakov*, 1986; *Mironenko et al.*, 1988; 1999; *Pashkovskiy*, 2002], not enough attention has been paid to preferential pathways of groundwater flow and migration.

The results of studies of preferential flow and migration pathways described below show that these pathways play an important role in the groundwater recharge and contamination. Their ignoring leads to a substantial underestimation of groundwater vulnerability. Therefore, it is very important to characterize in more detail the preferential pathways that we associate with PFMZs.

Under PFMZ we define the sites (in plane) or volumes (in space) of the geological medium, which, by their lithological composition, physicomechanical and geochemical properties, and other characteristics, reveal a higher aquife permeability and the corresponding transport of mobile substances (radionuclides, heavy metals, pesticides, etc.) with velocities essentially exceeding the background values. These zones can be of different size, origin, and evolution and can have varying water exchange characteristics and transport mechanisms for substances.

In groundwater resource assessments, insufficient attention to zones with abnormally high vertical water exchanges leads to underestimation of calculated forecast resources and is justified in definite conditions, for example, in view of diminishing expenses for research. However, when studying possible groundwater contamination and vulnerability and protectability to contaminants, such approaches may lead to significant underestimates of predicted risks as compared to their real values.

Groundwater Vulnerability: Chernobyl Nuclear Disaster, Monograph Number 69.
Edited by Boris Faybishenko and Thomas Nicholson.
© 2015 American Geophysical Union. Published 2015 by John Wiley & Sons, Inc.

Meanwhile, the presence of permeability anomalies in the structure of the geological environment is one of the main features controlled by structural-geodynamic heterogeneity and regularities of regional development as well as by the related exogenic processes.

Theoretical and experimental studies of physical regularities and mechanisms of preferential flow and transport in soils and rocks have been implemented for quite a long time. The concept of preferential flow of solutes in heterogeneous soils has been described first by *Lawes et al.* [1882]. Later on, *Deecke* [1906] published results of his observations of the dunes of the Darr peninsula in Germany concerning the formation of finger-like infiltration pathways in aeolian sands during heavy rainfall. Similar observations were made by *Gripp* [1961] in Germany and *Gees and Lyall* [1969] in Canada. *Raats* [1973] *and Phillip* [1975] explained these phenomena with unstable wetting front theories.

A series of studies during the last few decades were devoted to various types of preferential flow phenomena differentiated by their physical mechanisms. They include macropore flow in soils [*Bouma*, 1981; *Beven and Germann*, 1982; *Singh and Kanwar*, 1991], gravity-driven unstable flow [*Hill*, 1952], heterogeneity-driven flow [*Kung*, 1990], and oscillatory flow [*Prazak et al.*, 1992]. All the above-mentioned cases of preferential flow have relatively small scales ranging from 10^{-2} to $10\,\mathrm{m}$ [*Nieber*, 2001].

Helling and Gish [1991] used representations described earlier by *Landon* [1984] and proposed a general classification of the pore space and pore functions, including dimensions from 10^{-4} to $10^4\,\mathrm{mm}$. *Greenland* [1977] showed that pores with dimensions of less than $10^{-1}\,\mathrm{m}$ mainly have the function of joint capacity.

As a larger-scale process, depression-focused recharge should be mentioned with characteristic dimensions from 10 to $10^3\,\mathrm{m}$. This type of flow is well known to hydrogeologists and has been studied experimentally and theoretically [*Lissey*, 1971; *Nieber et al.*, 1993; *Gurdak et al.*, 2008; *Gerke et al.*, 2010]. An extending consideration of the groundwater focused and diffused recharge in various geological and landscape conditions applicable in the United States, including the karst groundwater recharge, is presented in the NRC report [2004].

Natural observations and experimental studies of the downward infiltration and migration of Chernobyl-born radionuclides in the depressions at natural test sites located in the Kyiv region and CEZ have been performed in Ukraine by Radioenvironmental Center, NASU, in the postaccident period [*Shestopalov and Bublias*, 2000, *Shestopalov et al.*, 2002].

It is worth noting that research into preferential infiltration and migration pathways was initiated in most cases by agricultural practice, for example, during research on groundwater contamination with pesticides and nitrates [*Shuford et al.*, 1977; *Parlange et al.*, 1988]. For this reason, most authors mainly described soils down to a depth of 0.3–1 m, and the preferential flow phenomenon was associated with heterogeneity of the soil pore space (macropore flow) and a heterogeneous finger-like moisture distribution in layered soils [*Glass et al.*, 1989; *Baker and Hillel*, 1990]. Some authors described preferential flow through the unsaturated zone to depths of 5–10 m and more [*Kung*, 1990; *Singh and Kanwar*, 1991]. Significant efforts were devoted to the theoretical and modeling aspects of the problem [*Hillel and Baker*, 1988; *Nieber et al.*, 1993; *Nieber*, 1996]. Extended researches were also performed in the United States of the preferential flow and migration mechanisms in solid fractured rocks [*Faybishenko et al.*, 2000, 2005].

Most of the contributions mentioned above described the preferential flow phenomena and their possible mechanisms in the detailed scale of soil sections and local sites. The regional aspects of the problem usually have not been studied.

On the other hand, in the regional assessments of groundwater vulnerability and protectability [*Aller et al.*, 1987; *Vrba and Zaporozec*, 1994; *Rosen*, 1994; *Rundquist et al.*, 1991; *Belousova and Galaktionova*, 1994; *Goldberg*, 1983; *Zektser*, 2001; *Pashkovsky*, 2002], the questions related to the preferential transport of groundwater contaminants were not taken into account.

3.2. PFMZ Classification and Occurrence

The processes mentioned above correspond to different-size zones of PFMZ in the geological medium. PFMZ classification in the geological environment may be done with respect to (1) genesis and corresponding degree of involvement of elements of the water exchange geosystem, (2) morphometry (surface, vertical section), (3) degree of activity, and (4) direction and character of development.

By morphometric characteristics, they may be classified into megazones (subregional) with dimensions ranging within 10^4–10^5 m: macrozones (10^3–10^4 m), mesozones (100–1000 m), microzones (10–100 m), nanozones (1–10 m), picozones (0.1–1 m), and femtozones (smaller than 0.1 m). The megazones are usually contoured in the large-scale studies. In the medium-scale and detailed-scale studies, the smaller forms, such as macrozones, mesozones, and microzones with dimensions from 10 to 10^4 m, can be contoured and studied in detail.

Nanozones and picozones (and smaller) are usually not manifested or are only weakly distinguished in the relief. Their heterogeneity is determined by the character of sedimentogenesis and diagenesis of rocks and soils and the influence of geobiocenoses. Following *Greenland* [1977], one can accept that they perform the function of a joint capacity of the geological medium and participate in formation of its entire background permeability and infiltration field. Their anomaly properties are reasonable to account for in consideration of small-size fields (from meters to tens of meters).

By their genesis, the PFMZ may be classified into two large groups: exogenic and endogenic. The first ones relate to peculiarities of erosion, glacial processes, karst, suffusion processes, influence of biocenoses and technogenic disturbances on the structure of the soil layer, unsaturated zone, and even part of the saturated zone. The second ones are determined by distribution of the geodynamic stresses and degassing processes in Earth's crust, geochemical processes, and the lithofacial peculiarities of deposits. Because of the long-term influence of endogenic factors, they often lead to the appearance and imposition of exogenous processes on the endogenous formation of the PFMZ. As a result, in many cases it is possible to distinguish a third group of zones with a mixed genesis. These are formed under the influences of internal and external forces but are mainly due to endogenic stresses and processes. The presence of endogenic, geodynamic, and gas-related factors in the formation of PFMZ determines their maximum development depth (hundreds and thousands of meters).

The lithofacial and geochemical factors of PFMZ formation usually determine their lower depths (mainly meters and tens of meters). Even smaller depths are characteristic of the exogenic PFMZ (usually within 1 or a few meters).

Due to the presence of morphological peculiarities in the section, PFMZ may be classified in the plane as linear, oval and round, and crescent-like and combined and in the section as window-like, vein-like, fractured, elementary (filled or not filled), etc.

Based on the infiltration and migration activity, PFMZ can be subdivided into hyperactive, with more than a tenfold increase of infiltration and migration processes as compared to background sites; very active, with an increase of these processes of 5–10 times as compared to background values; medium active (2–5 times higher than background); and low active (less than 2 times higher than the background). It is worth noting that the infiltration and the migration activity, depending on the deposit composition and migrant type, may not correlate.

According to the involvement of the elements of the hydrogeological system (soil, unsaturated zone, upper groundwater, deep confined aquifers, and aquitards), one can distinguish the following PFMZ types:
1. Thorough, with a complete or high involvement of the main water exchange geosystem elements
2. Mesodeep, with the involvement of soils, the unsaturated zone, and one or two aquifers
3. Subsurface (aeration), with the involvement of soils, the unsaturated zone, and possibly a part of the upper aquifer (in the case of a thin unsaturated zone)
4. Soil, located within the soil cover
   By evolution, one can distinguish stable, progressing, and degrading PFMZ.
   In solid and semisolid rock massifs, the macrozones, nanozones, and picozones of increased permeability related with fracturing are also known. However, their role as zones of preferential flow and mass exchange is usually assessed only integrally.

Special importance for the PFMZ phenomena that represent the micro-PFMZ and meso-PFMZ is usually related to the depression-type relief. For this reason, more detailed experimental studies have been implemented mainly of the closed-type relief depressions, for which the following classification can be given.

By morphometric indicators, the depressions can be classified according to their dimension (maximum cross-sectional length) and maximum depth. By dimension, the depressions are subdivided into small (length 10–50 m), average (50–100 m), large (100–1000 m), and very large (over 1000 m); by depth, they are classified as shallow (with a depth less than 0.5 m), average (0.5–2 m), deep (2–3 m), and very deep (deeper than 3 m).

In addition to these depression types, closed valleys also occur with dimensions of 1 km and more. The borders of such valleys have gentle slopes (below 1°) and weak surface runoff.

Based on the character of moistening, the depressions of the moderate-humid climatic zone can be classified into dry, damp (with periodic wetting), wet (during the most part of a year), and flooded.

The distribution of these types of depressions between the characteristic landscapes also obeys a definite regularity. Dry depressions mostly occur on sandy ridges and partially on water-divide plateaus. Damp depressions occur practically in all landscape types; however, they are most frequent at sandur (glacial outwash) plains in the CEZ. Wet depressions mostly occur within first floodplain terraces and elevated floodplain areas, and the flooded depressions frequently occur in the pass-through valleys, in low floodplains, and within rear joints of first floodplain terraces.

It is necessary to draw attention to the fact that the prevalence of PFMZ as depressions and linear zones is not just a feature of the territory in the vicinity of the Chernobyl nuclear power plant. They are widespread on the planet, and hence the importance of taking them into account in the assessment of groundwater vulnerability is relevant to many areas.

To assess the characteristics of PFMZ development on different continents, we conducted a sample survey of satellite images of individual plain areas These images, which are available through the Google Planet Earth program, show that in different landscape complexes variable depression forms exist (round, oval, linear, closed, and open) with varying area and activity of individual forms.

Practically in all continents the areas with depression microrelief can be identified. Most of them occur within the large-plain areas, in particular, plains in the river basins of the Dnieper, Volga, Elbe, Vistula, San Francisco (Brazil), Parana (Argentina), and Murray (Australia).

Of particular interest are the data on the PFMZ morphology on the North American continent. An analysis of the satellite images of the area identified various combinations of round depressions and linear forms. Here the areas are represented with mainly primary fracturing of rocks, which on the surface images is expressed with thin lines; fractures that have undergone secondary effects;

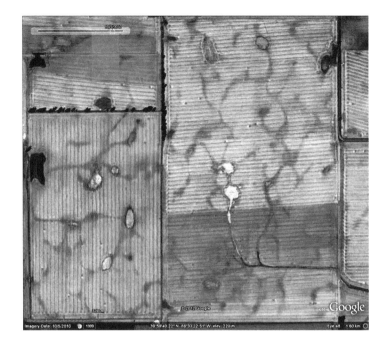

Figure 3.1 United States, Illinois, Monticello, 30 km SE from Klinton NPP.

depressions within the grid fracture systems (Figure 3.1); mixed-type depression forms; depression microrelief with low primary fracturing (Figure 3.2); and depression microrelief with no visible primary fracturing.

Sample estimate of PFMZ was performed also by images of the European continent, especially in Ukraine. In some areas their number exceeds 200 per square kilometer. By morphological types, the rounded and oval depressions dominate in the region, ranging in size from tens of meters to several kilometers (Figures 3.3–3.5).

Based on the assessment of regional peculiarities development, it was found that the number of depression forms per definite area decreases with the degree of area dissection.

In western Europe the depressions are also widely developed. Most depressions are associated with plains of large river basins. However, because of the often high water table, their manifestation in the relief may not be very significant. The clearest manifestation in the relief that the depressions have on elevated areas is shown in Figure 3.6.

As we see, even a quick review of the depression forms shows their worldwide occurrence. Their role in groundwater vulnerability and protectability requires special account and study.

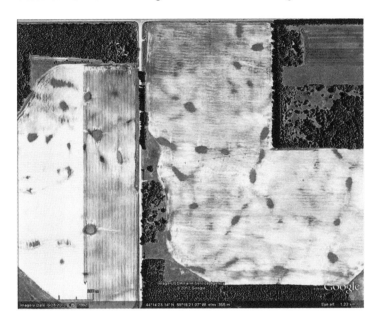

Figure 3.2 United States, Wisconsin, 20 km N from Wautoma.

Figure 3.3 Ukraine, CEZ, 20 km W from Chernobyl NPP.

Figure 3.4 Ukraine, Chernigiv oblast, Yablunivka, 18 km SW from Pryluki (E from CEZ).

Figure 3.5 Ukraine, Poltava oblast, Dyachenki, 60 km SW from Poltava (SE from CEZ).

Figure 3.6 Germany, Mecklenburg-Vorpommern, Granzin.

3.3. Methodological Approaches of PFMZ Study

The study of depression-type PFMZ was conducted at six sites in the CEZ and the experimental site "Lyutezh," 30 km north of Kyiv. In addition, some PFMZ components have been studied at pilot sites of Chernihiv, Mykolaiv, and Odessa regions.

As a methodological framework, the landscape and geological principles of study were accepted. All methods have been specially selected and tested at the experimental site. The objective of the PFMZ study was to obtain a wide range of factual data that can be grouped according to the following: (1) the nature of regional development of depression forms, (2) the external and internal structures of the depressions, (3) the morphogenetic reconstruction, (4) factors affecting PFMZ formation and development, and (5) current processes and modeling of PFMZ changes in comparison with the background sites.

For the regional assessment of the PFMZ features, satellite images, aerial photographs, and topographic maps were used. In the key sites the results of the off-site analysis were checked.

The detailed study of depressions allows determination of the following:

1. Degree of activity of individual morphological components or separate depression forms by the nature of the slopes (concentration, dispersion of elevation isolines contour lines)

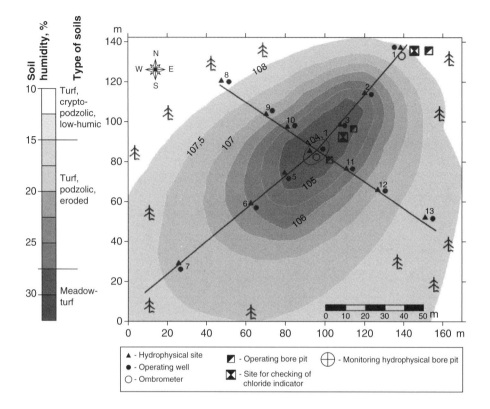

Figure 3.7 Scheme of a surface depression with technical elements of its research.

2. Places with the most active or passive geodynamic processes by the degree of stability of the soil and vegetation cover
3. Stage of depression development (ancient active, young active, ancient passive) by the character nature of the depression bottom (convex, concave, flat) and changes in the thickness of stratigraphic horizons (as compared to the background)
4. Character of rock watering at certain stages of their development by the traces of rock hydromorphism at some horizons
5. Temperature and humidity conditions by traces of ancient cryogenesis
6. Activity of geochemical processes by the size and number of tumors

Most of the depressions are oval, often extended along the long axis (Figure 3.7), with asymmetrical sides. On steep slopes disturbance of soil and vegetation often takes place. From background areas to the central parts of the depressions, change in the genetic varieties of soils occurs from zonal to azonal transformed by the hydromorphism and the specific nature of the development and humification of biological objects.

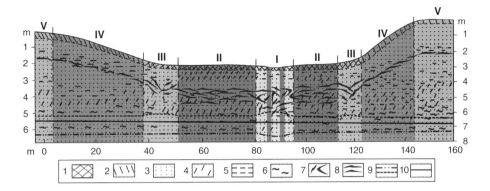

Figure 3.8 Complex scheme of geological section (160 m wide) across a characteristic depression (Lyutezh observation plot, northern part of Kyiv region) with lithostratigraphic elements and zones of different hydrodynamic activity: (1) sod-meadow loamy sandy soil; (2) sod-podzolic sandy soil; (3) fine-grained sands with seams of fractions: (4) fine dusty; (5) coarse dusty; (6) clayey; (7, 8) ferruginous seams; (9) groundwater level in winter (double-dotted dashed) and summer (dashed) periods; (10) annual average groundwater level. Zones: I, central active; II, central passive; III, near-border active; IV, slope passive; V, background.

The most common internal characteristics of the central parts of depressions (Figure 3.8) are changes of the thickness of stratigraphic horizons, high hydromorphism, leaching of rocks, a high level of rock disturbance by ancient cryogenesis (mainly in the form of wedges and viens), and the formation of various tumors and local geochemical barriers.

In the central part of the depressions in places of vertical cryogenic pseudomorphs (wedges, veins, cauldrons, involutions), the increased flow of water and solutes (and sometimes of suspended fine fraction) is observed. At the boundaries of the lithological differences, the tumors are usually found in the form of films, "beans," "cranes," tubes, peels, layers, lenses, etc., made mostly of iron sesquioxide nodules (in humid conditions), which often serve as a geochemical barrier to dissolved and suspended phases.

For the structure of the transition zone (closer to the sides), a characteristic is the smaller impact of hydromorphism, cryogenesis, and the chemical and physicochemical transformations of rocks. In the sections there is a clear differentiation of the stratigraphic and lithologic horizons, traces of paleocryogenesis presented as small cracks, thin layering, and low-graded textures. In the near-slope zone, the attenuation of geodynamic processes is seen: the small effect of hydromorphism and faint traces of the geochemical and cryogenic transformation of rocks.

In rocks of central depressions, a lack of readily soluble minerals, degradation of the macroaggregate and macroporous structure, increasing density of rocks (from the sides to the center), reduced permeability in undisturbed places,

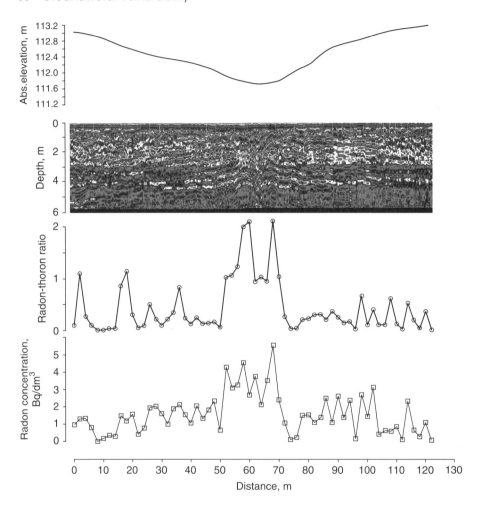

Figure 3.9 Relief elevation and results of GPR and emanation profiling across the depression (observation plot Stary Shepelichi, CEZ). (*For color detail please see color plate section.*)

and increased permeability in places disturbed by cryodynamic and thermodynamic processes are usually observed. These signs indicate relatively high dynamics of liquid and solid phases of the geological medium and increased activity of geochemical processes within the central parts of active depression forms.

In the study of the structural elements and the whole PFMZ structure of the cover deposits (to a depth of 10 m), the ground-penetrating radar (GPR) SIR-2 with a 300 MHz antenna was used. In the result, the sections were obtained with detailed reflection of horizontal and vertical heterogeneity (Figure 3.9).

The characteristic sizes of unconsolidated zones in rocks and some anomalous properties (degree of fragmentation) are well recorded by *emanation profiling*, which consists in identifying abnormal manifestations of the radioactive gases thoron (^{220}Tn), radon (^{222}Rn), and sometimes carbon dioxide (CO_2).

For the measurement of radon and thoron emanations in soil air, the alpha-beta analyser and gamma-spectrometer NC-4826 were used. The emanation profiling data represent graphs showing space or space-time variations in the manifestations of radon and thoron emanation (Figure 3.9). The intensity of anomalies is determined by the ratio of the emanation field in soil to the background emanation value. In the active zones of depressions, the emanations intensity often is 2 and more times higher than at the corresponding background area. By their composition the emanations are classified as radon, thoron, and mixed-type radon-thoron and thoron-radon. As the corresponding emanation characteristic, which can also be related to the intensity of the anomalous field, the radon-to-thoron ratio is often calculated. The distinguishing feature for anomalies is that at background sites the statistical error of emanation measurements for radon and thoron is high (reaches 100%), though in anomalous zones it is significantly lower (about 20–30%).

Along with GPR and emanation, seismic profiling was used on the same experimental plot to build the signal diagrams along profiles passing across the studied depression through its background and anomalous zones. The seismic signal was generated using the shock method. The reflected signal was detected by sensors placed before the point of excitation and was recorded on the field computer. On the resulting signal processing diagram, the zones of the reflected signal intensity represent a certain structure of the depression "body" down to a depth of 50 m. (Figure 3.10). Interpretation of the selected fields with different signal levels was carried out on the basis of data taken from the reference wells. Analysis of these materials showed that in the central part of the depression and partly on the slopes there are areas with relatively higher deposit humidity. These data are in good agreement with the above results of GPR and emanation profiling.

In the result of the implemented complex of geophysical studies [*Shestopalov and Bublias*, 2000], it was found that in all the studied depressions the zones exist characterized with the different state of the geological environment as compared to the background areas.

In determining the quality and quantity of the most mobile component of the geological environment (the liquid phase, containing dissolved and suspended substances), hydrophysical and chemical methods were used, which allowed control of the water-salt regime, flow rate, and pore water balance [*Shestopalov*, 2001].

Determinations of the flow rate and volume of infiltrating water in the vadose zone were carried out using ceramic gauges for the pore water suction pressure in the unsaturated zone at different depths (usually a 10 cm interval to depth 0.5 m and deeper and a 1 m interval to depth 3 m). Necessary parameters

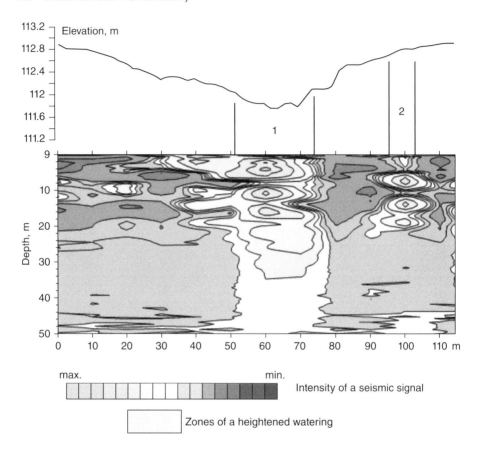

Figure 3.10 Seismic profile diagram obtained across the depression at observation plot Stary Shepelichi, CEZ: (1) central active zone; (2) slope active zone. (*For color detail please see color plate section.*)

such as the moisture transfer coefficient and the volumetric moisture content, depending on the suction pressure, were determined in the laboratory. The gauges and observation wells were installed in sites identified by the geophysical methods described above in places of high infiltration (active zones), low infiltration (passive zones), and medium infiltration (background areas) (see Figure 3.8).

    To study the characteristics of migration of radionuclides, both field and laboratory methods were used. Field works included the measurement of the exposure dose of radiation with the gamma-field radiometer. To identify differences in radionuclide migration, the soil sampling for radioactivity was conducted in different morphological elements of depressions. Soil samples were collected on the horizontal and vertical profiles. Measurements of radionuclide concentration were carried out for groundwater samples taken from the regime wells.

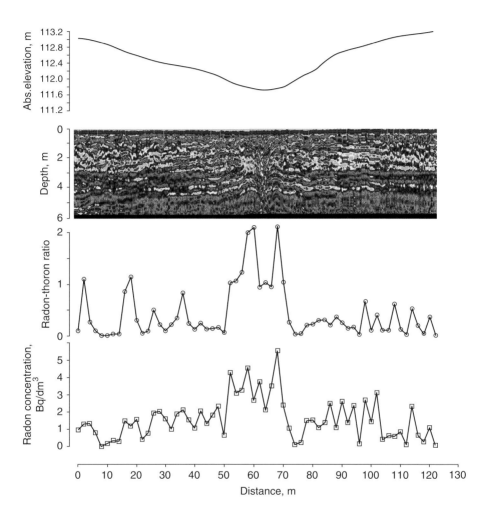

Figure 3.9 Relief elevation and results of GPR and emanation profiling across the depression (observation plot Stary Shepelichi, CEZ).

Groundwater Vulnerability: Chernobyl Nuclear Disaster, Monograph Number 69.
Edited by Boris Faybishenko and Thomas Nicholson.
© 2015 American Geophysical Union. Published 2015 by John Wiley & Sons, Inc.

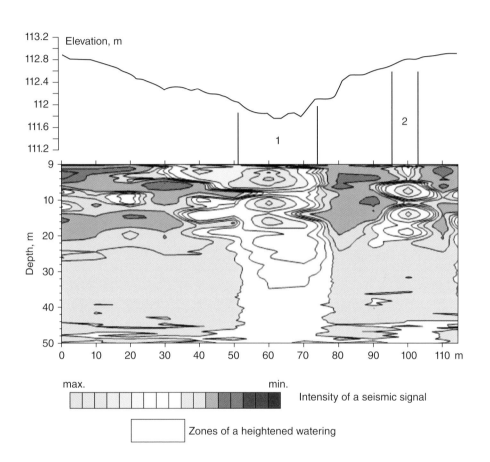

Figure 3.10 Seismic profile diagram obtained across the depression at observation plot Stary Shepelichi, CEZ: (1) central active zone; (2) slope active zone.

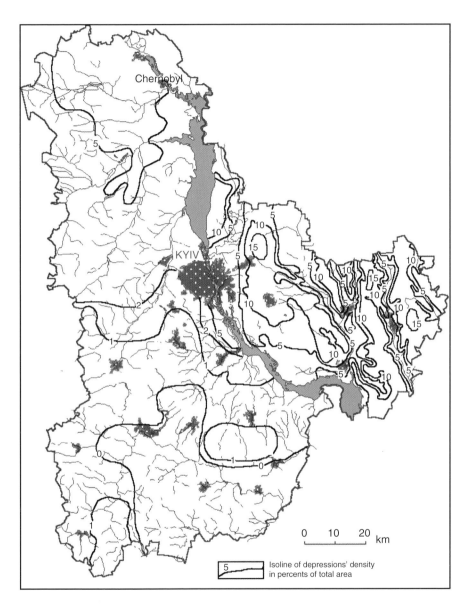

Figure 5.4 Distribution scheme of depression densities (percent of area covered by depressions) for the Dnieper basin (Kyiv region) obtained from a cartographic analysis in scale 1:50,000.

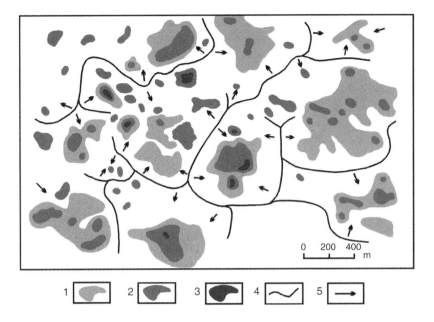

Figure 3.11 Scheme of depression occurrence at typical site of a sandur plain in the CEZ with indication of absolute elevation, depression depth, watershed dividing lines, and surface runoff directions: (1) depressions with depth to 0.5 m; (2) depressions with depth to 1.0 m; (3) depressions with depth to 1.5 m; (4) watershed dividing lines; (5) surface water flow directions.

3.4. Indicators of PFMZ Activity in Depressions

Detailed experimental studies were carried out mainly of closed-type depression morphosculptures for the reason that in the plain areas their contribution to the vertical migration component is much higher than that of the poorly defined linear forms. The role of linear open forms in the vertical mass transport increases significantly in the dissected relief with absence or poor development of closed depressions.

Results of field studies and experiments implemented in the study area of CEZ and the Kyiv region showed that mesoscale and microscale PFMZs represented by closed depressions exert a significant influence on the intensity of the vertical infiltration and transport of radionuclides and other contaminants from the surface into groundwater.

The depressions serve as the bases of local erosion and runoff and form local watersheds of surface waters that converge on their central lowest parts (Figure 3.11).

It was found that in the CEZ depression systems may capture up to 60–80% of surface runoff of plain water-divide areas [*Shestopalov*, 2001]. This surface runoff mainly forms during periods of spring snow melting and intensive rainfalls.

The infiltration value was calculated based on the known intensity of precipitation, surface runoff, evaporation, and transpiration. These calculations have shown that at the background areas the infiltration reaching the upper aquifer equals about 100–210 mm/year, and in the depression it ranges from 500 to 700 mm/ year. In wet years this value may reach 1000 mm/year or more, depending on the nature and amount of precipitation, type of depression, and its catchment area [*Shestopalov et al.*, 1997, *Shestopalov and Bublias*, 2000]. Study of groundwater infiltration at the Lyutezh experimental site (1998–2002) with a chlorine indicator showed that in the depression it ranges from 450 to 550 mm/year, while in the background site it has lower values, from 200 to 240 mm/year [*Shestopalov*, 2001].

During periods of heavy rainfall, in the central part of the depressions the spreading cupola formed up to 20 cm high or more, and in dry periods depression cones with a relative depth of 15–20 cm are often formed. These phenomena indicate, on the one hand, that large volumes of water pass through the center of the depression into the groundwater and, on the other, that possible active water exchange occurs of the upper aquifer with the deeper ones.

Another typical example is the data records for infiltration in the vadose zone of the observation plot "Stary Shepelichi" for the high-infiltration period in 1995 (Figure 3.12). In this territory the annual precipitation comprised 675 mm; during the summer it was about 480 mm. For the warm period of the year the average infiltration in the central part of the depression was 803 mm, and in the background area it was just about 85 mm. During this period, in the central part of the depression a spreading cupola was observed of the groundwater table with a relative height of 10–20 cm.

An important indicator of increased mass exchange in the depression-related PFMZ is increased activity of the geochemical processes, which has a definite reflection on the chemical composition of groundwater and the mineralogical composition of rocks. In rocks of the depression central parts, the absence of easily soluble minerals, degradation of the rock structure by leaching, and destruction of colloidal shells of elementary particles often take place (Figure 3.13).

The groundwater chemical composition analysis of the central anomalous zone of the depression at the Liutezh experimental plot in the Kyiv region and corresponding background zone (Figure 3.14) show that the groundwater mineralization in the PFMZ is significantly lower than in the background areas. This observed mineralization behavior evidences the higher washout degree of rocks and deposits in the central zone of the depression in conditions of increased infiltration and dilution.

Based on the above results, one can suppose that the rate of geochemical processes and migration of natural water solutions in the active zones of depressions are several times higher than in the background areas.

The study of radioactive contamination of rocks in the unsaturated zone of depressions in the CEZ during 1987–2000 showed the fast transport of

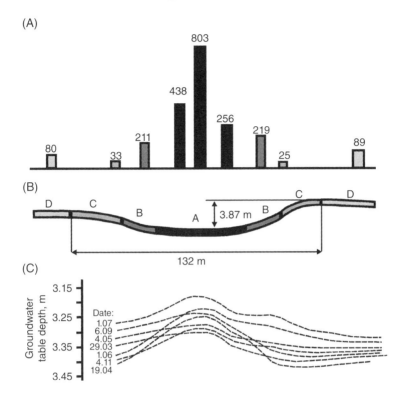

Figure 3.12 Average annual infiltration and variations of groundwater levels in the depression at observation plot Stary Shepelichi (CEZ): (A) infiltration rate (mm/year); (B) land surface profile; (C) groundwater level variation. Depression concentric zones: (**A**) central active; (**B**) intermediate; (**C**) slope; (**D**) background.

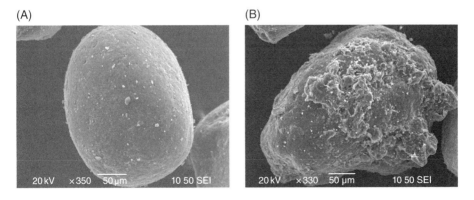

Figure 3.13 Morphology of elementary rock particles by electron microscopy: (left) background zone of a depression; (right) active zone (Liutezh experimental plot).

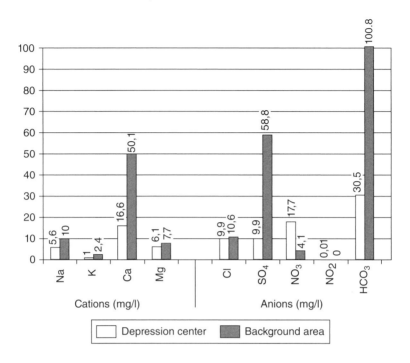

Figure 3.14 Comparison of the chemical composition of groundwater of the upper aquifer in the depression central active and background zones by data from the observation plot Liutezh (Kyiv region, sampling performed in October 2006).

radionuclides from the watershed area to the central depression part and their further penetration into groundwater. In the central active parts of some depressions the Chernobyl-born radionuclides ^{137}Cs and ^{90}Sr were found in rock samples even at depths significantly below the groundwater table. For example, at the Veresok site in the CEZ the radionuclides were found at a depth of 17m, and the groundwater table depth was 4m. At the corresponding background area the depth of radionuclide penetration did not exceed 6m. Generally, within the Kyiv region the radionuclides were found at depths of 100m and deeper [*Shestopalov*, 2001].

The surface postaccident contamination with radionuclides (^{90}Sr and ^{137}Cs) in the CEZ reached values up to 10^6 Bq/m^2, and in the Kyiv region contamination was up to 10^4 Bq/m^2. The comparison was performed of the intensities of vertical radionuclide transport in the central active and background parts of depressions. For this purpose the rock samples were taken in pits to a depth of 1m and analyzed in the radiochemical laboratory for radionuclide concentrations. The obtained vertical concentration distributions for ^{90}Sr (depression of "Veresok" observation plot, CEZ) and ^{137}Cs ("Liutezh" observation plot, Kyiv region) for depression central zones and corresponding background sites are shown on

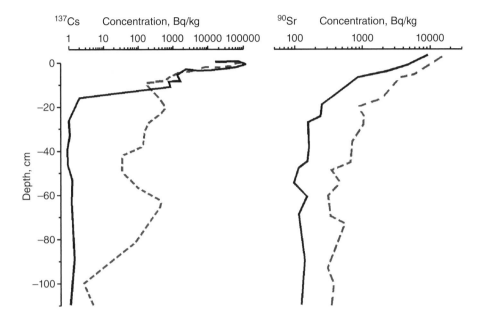

Figure 3.15 Vertical distribution of radionuclide concentrations in rock samples: ^{90}Sr (observation plot Veresok, CEZ) and ^{137}Cs (observation plot Liutezh, Kyiv region) according to postaccident data. Dashed line, depression center; solid line, background area.

Figure 3.15. The observed difference of radionuclide activities in the central parts of depressions and background sites reached 2 orders of magnitude for ^{137}Cs and 1 order of magnitude for ^{90}Sr. This provides evidence of a higher relative intensity of vertical radionuclide transport in the PFMZ.

3.5. Preliminary Evaluations of PFMZ Influence on Upper Groundwater

Let us consider a closed watershed area represented conventionally by two landscape types: background and depression-related PFMZ. Let $F_s = F_b + F_a$ (m²) be the total watershed area; F_b and F_a be areas of background and depression PFMZ sites, respectively; and W_b (m/day), C_b and W_a, C_a be average infiltration and contaminant concentration in groundwater at the groundwater table depth at background sites and depression PFMZ sites, respectively. Then, the balance equations determining the flow and contaminant mass balance within the watershed can be written in the forms

$$W_s F_s = W_b F_b + W_a F_a, \tag{3.1}$$

$$C_s W_s F_s = C_b W_b F_b + C_a W_a F_a, \tag{3.2}$$

where W_s is average infiltration recharge over the whole watershed and C_s is the resultant average (over the whole watershed) contaminant concentration. According to these equations, determined by C_s is the formula

$$C_s = \frac{C_a W_a F_a + C_b W_b F_b}{W_a F_a + W_b F_b}. \tag{3.3}$$

The share of depressions in the total groundwater contamination is determined by the ratio

$$\eta = C_a W_a F_a / C_s W_s F_s. \tag{3.4}$$

Using the available data on the infiltration intensity and vertical migration of ^{137}Cs from the contaminated surface to the groundwater table depth for a typical depression in the CEZ and $\sigma = F_a/F_s = 0.1$ (observation plot "Stary Shepelichi"), an approximate assessment of the input of depression-related PFMZ in the total contamination of upper groundwater was obtained [*Shestopalov et al.*, 2006] according to formula (3.4), $\eta \approx 0.82$. Hence, we came to the conclusion that the share of anomalous depressions in the total radionuclide contamination of the upper groundwater by ^{137}Cs may reach 80% and more. Also, the relative share of depressions in groundwater contamination, η, was calculated as a function of groundwater table depth for three values of the areal share occupied by depressions, $\sigma = F_a/F_s$: 1%, 5%, and 10%. The corresponding plots are shown on Figure 3.16.

It is seen from Figure 3.16 that the relative share of depressions in the total influx of the contaminant with the infiltration to the upper groundwater, according to formula (3.4), increases from 3–30% to 80–100% with increasing groundwater table depth. In areas with a groundwater table depth greater than 15 m, the penetration of ^{137}Cs into the upper groundwater aquifer proceeds mainly on account of depression-related PFMZ (by 90% or more).

To analyze the joint effect of two types of PFMZs (depressions and lineaments), we have considered a pilot area of 149 ha located in the southern part of the Kyiv region [Rokitne district, west from Bakumivka village (Figure 3.17)]. Figure 3.17A is a satellite image of the area (data Cnes/Spot Image, 2011) with typical depressions and lineaments (elongated linear and curvilinear PFMZ). Figure 3.17B shows the result of gridding the depressions (dark) and lineaments (light) with a square grid spacing of 10 m.

According to our preliminary assessment for this area, it was supposed that the depression zones received 2 times increased infiltration and predicted relative concentration of the contaminant (^{137}Cs) in the upper aquifer as compared to background area, and the lineament zones received 1.5 times increased infiltration and contaminant concentration. As a result of counting areas of the depression-related and lineament-related PFMZs, the following data were obtained: the

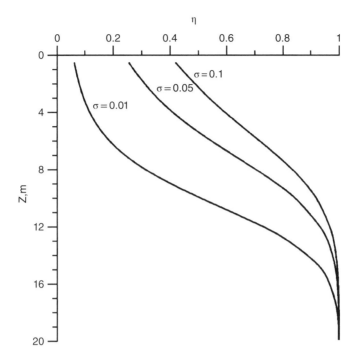

Figure 3.16 Plots of the relative share of depressions in the upper groundwater contamination with ^{137}Cs η on the groundwater table depth Z for different values of the areal share of depressions, $\sigma = F_a/F_s$.

area of depressions is 15.24 ha (10.23%), and the area of lineaments is 12.69 ha (8.52%). The respective background area is 121.07 ha (81.25%).

Performing the balance calculations in the same way as given by equation (3.1) but for two active zone types (depressions and lineaments), we obtained 20.4% share of depressions in the average infiltration over the plot, 12.8% share of lineaments, and 66.8% share of background areas. Using the formula of balance averaging for concentrations similar to (3.3) and taking the relative groundwater pollutant concentration in the background as 1 and the corresponding relative concentration in depressions as 2 and in lineaments as 1.5 (according to our preliminary estimate), we obtain the assessed share of depressions in the general pollution of upper aquifer groundwater area as 41%, the share of lineaments as 19%, and the corresponding share of background areas as 40%. Thus, according to this preliminary result, the depression and lineaments, despite their relatively small proportion of the total area of the site, may be responsible for 60% of total income of the pollutant into groundwater.

It is obvious that the preliminary assessments given above for the relative influence of one and two types of PFMZ can be considered only as a first approximation in the assessment of the role of anomalous landscapes in the infiltration

(A)

(B)

Figure 3.17 Satellite image of the area for assessment of the influence of two types of PFMZ (depressions and lineaments) (A) and their gridding (B) by square elements.

recharge and contamination of upper groundwater. It can be treated as a preliminary stage of PFMZ account in assessing groundwater vulnerability and protectability for the given area.

In general, the described concept of the role of different-type PFMZs such as depressions and lineaments in the formation of groundwater recharge and fast migration of potential surface contaminants into the groundwater can be represented as shown in Figure 3.18.

3.6. Practical Importance of PFMZ

The relevance of problems related to PFMZs extends beyond the local area in which detailed studies were implemented. It is global because the microstructure of the geological environment is typical for almost all regions of Earth. Let us consider some aspects of their practical importance:

1. In the hydrogeological aspect, the importance of depression-related PFMZ is determined by their possible significant contribution to the formation of

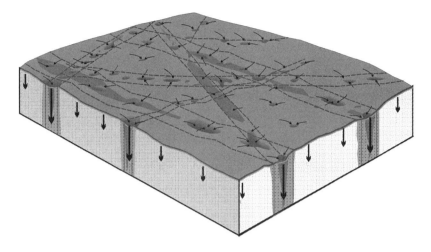

Figure 3.18 Scheme of groundwater recharge and contaminant transport by PFMZ related with relief depressions and lineaments.

natural resources and operating reserves of groundwater while increasing their vulnerability to pollution.

2. In areas with a lack of significant practical use of aquifers in human-caused accidents leading to significant surface contamination, the depression-related PFMZs play a positive environmental role, which is to accelerate the deepening of surface contaminants and their immobilization in the deeper geological environment. In this case, the depressions accelerate the remediation of contaminated areas.

3. Thus, evaluation of the quality of land for the production of organic agricultural products depends on the degree of area infestation by anomalous depressions. It has been established that the coefficient of accumulation of radioactive elements and heavy metals in plants growing in the depressions is much higher than in the background areas. The amount of agricultural product also has a direct link to depression-related PFMZs. According to official data, from 35 to 40% of agricultural product is lost due to these depressions. Regarding the fact that only the depression forms (others are not counted) cover about 2.5 million hectares of arable land of Ukraine, it is clear that they bring huge losses for the state.

    On irrigated lands the depressions hinder irrigation. Therefore, in some areas of the steppes of Ukraine attempts were made to align them by filling with ground. However, after a few years new depressions formed of even larger sizes.

4. The depression zones bring many disadvantages to builders, because the grounds within them have less favorable geotechnical properties and their

characteristics increased dynamic processes. When a construction site of various objects locates in these areas, this usually leads to an increase in the estimated cost at all stages: research, design, construction, and operation. Of particular importance are the specific questions of the anomalous zones developed for selection of areas for the construction of high-risk objects such as chemical plants, nuclear power plants, and repositories of radioactive waste, where the level of stability and security of the geological environment are of great importance.

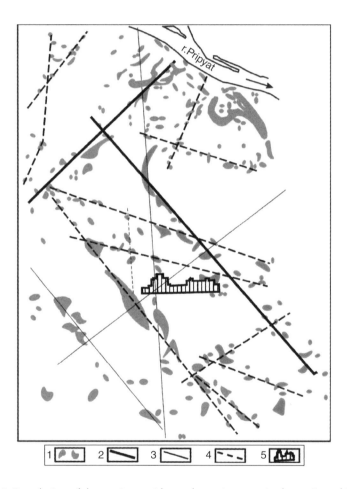

Figure 3.19 Correlation of depressions with geodynamic zones in the region of Chernobyl NPP location: (1) depressions; (2) zones of tectonic breaks revealed by geophysical methods; (3) geodynamic zones revealed by data of aerospace images decoding; (4) geodynamic zones associated with linear groups of depressions; (5) Chernobyl NPP.

Engineering and building structures located within the geodynamic zones are subject not only to increased mechanical stress but also to chemical and microbiological destruction. In particular, the incidence of iron bacteria is several times higher in these areas than in the background ones. This significantly reduces the service life of steel and concrete structures. In active geodynamic zones oxidation and reduction reactions create a hostile environment for a variety of building materials.

Violations of foundations and structures, buried in sediments under these conditions, lead to increasing risks of groundwater contamination since they create additional potential sources of contamination at hazardous chemical and nuclear facilities, such as NPPs.

Special efforts have been implemented to identify PFMZs in the Chernobyl NPP area. Within its operation site and the adjacent areas, based on aerial photographs and topographic maps, geodynamic zones that appear in the current landscape (topography, soils, vegetation) were identified. It is established that the location site of the NPP is crossed by several active geodynamic zones (Figure 3.19).

Important is the fact that, during the Chernobyl NPP construction, its industrial site was covered with a layer of sand and leveled. However, after 16 years, the depressions again appeared in relief. Some of them have achieved a significant depth, indicating the activity of the processes of migration and matter removal from these areas through the geological environment.

The survey of the CEZ area (see Figures 3.3, 3.11, and 3.19) and the locations of NPPs in other countries (see Figure 3.1) show that the round and linear depression forms are of frequent occurrence in these areas.

These examples show that the geodynamic anomalies presented by depressions and linear forms of relief have a deep nature of formation and development, and this conclusion should be implemented in practice.

4. METHODOLOGY OF GROUNDWATER VULNERABILITY AND PROTECTABILITY ASSESSMENT

4.1. General Consideration

The existing groundwater vulnerability assessments considered in Chapter 1 in most cases do not account for preferential flow and migration related to PFMZ in the unsaturated zone and, moreover, in the saturated zone of aquifers and aquitards. Special methods accounting for the preferential flow were implemented using the index and rating methods in a case of karst areas. Meanwhile, the influence of different-scale flow and migration heterogeneities in the upper geological environment (from pores and fractures in the soil and in the unsaturated zone to depressions and lineaments of relief and tectonic dislocations) on the predicted groundwater contamination appears to be very significant. This fact, as was considered above, becomes evident from the results of studies of real groundwater contamination by Chernobyl-related radionuclides over wide areas. Consequently, the most important task is the development and improvement of the groundwater vulnerability assessment methods with respect to accounting for PFMZs.

From our viewpoint, methods of mathematical modeling based on representations of physical processes related to contaminant transport in the groundwater and its balance assessment are the most appropriate among the existing methods of groundwater vulnerability assessments when accounting for PFMZs. The modeling method described in Chapter 1 using the 3D model (1.14)–(1.17) of groundwater flow and contaminant transport by advection, dispersion, and sorption is sufficient for the characterization of groundwater vulnerability, taking into account PFMZs. The difficulty, however, appears in the process of the realization of the method, which requires the use of great volumes of data for the parameters of geological media in particular, hydraulic conductivity, retention capacity, and lithological heterogeneities to account for different-scale PFMZs. The information about the spatial distribution of flow and transport parameters of the geological media, such as conductivities, sorption (distribution) coefficients,

Groundwater Vulnerability: Chernobyl Nuclear Disaster, Monograph Number 69.
Edited by Boris Faybishenko and Thomas Nicholson.

and porosities is almost always insufficient for obtaining a reliable solution of a 3D flow transport modeling, even without accounting for PFMZs.

One way of overcoming this situation is the solution of inverse problems to assess missing flow transport parameters of the medium using all available data (e.g., literature) and results of real groundwater contamination studies (in particular, radionuclides of the Chernobyl origin) at experimental test sites and areas characterized by different-scale PFMZ occurrences.

Another effective technique for simplifying the groundwater vulnerability assessment model (taking PFMZs into account) is combining the method of hydrogeological zoning described in Chapter 1 with the corresponding 1D modeling and comparing with the observation results. This allows (with a certain degree of convenience) to reduce the model dimension from 3D to 1D and to con-sider, for example, 1D vertical transport from the contaminated surface to the groundwater aquifer. We used this approach in the development of the proposed methodology for assessing groundwater vulnerability to the Chernobyl-born radionuclide ^{137}Cs within the Kyiv region area of the Dnieper basin [*Shestopalov et al.*, 2006]. Let us consider the methodology in more detail.

The methodology is based on 1D mathematical modeling of the vertical contaminant migration from the soil surface (in the upper groundwater assessment) or from a given (previously assessed) aquifer into the deeper aquifer. The downward transport of a soluble contaminant in the unsaturated and satu-rated zones should be propagated by advection, dispersion, and sorption mecha-nisms. We neglect the molecular diffusion because its influence, according to numerous data literature, is insignificant in the scale of the assessment considered [*Bochever and Oradovskaya*, 1972; *Lukner and Shestakov*, 1986]. The model allows for the determination of the vertical contaminant concentration distribution at a given time $C(z,t)$ in the infiltrating water at depth z during a given time period $0 < t < t*$ from the initial contamination fallout and known initial (at $t = 0$) value of contaminant concentration C_0 in the liquid phase at the top model boundary (surface) $z = 0$.

The 1D modeling approach puts a definite restriction on the groundwater vulnerability assessment; that is, the vulnerability of a given aquifer is assessed with respect to the penetration of a contaminant into the aquifer with the infil-trating water through the overlying covering deposits. The assessment of the pre-dicted contaminant concentration is performed for a definite characteristic depth $z*$, which can be chosen either on the aquifer surface (unconfined groundwater table or upper boundary of a confined aquifer) or below the aquifer, depending on whether the protection properties of water-bearing rocks of the assessed aquifer itself are considered or not.

The forecast time $t*$ for the modeling of the transient contaminant concentration distribution $C(z, t)$ should be comparable with the lifetime of the given contaminant. For a radionuclide, the characteristic time period is represented by its half-life period. The forecast period $t*$ is assessed a priori in such a way that during this period a significant (or even maximum) concentration at the given assessment depth can be reached.

The *attainment time* necessary for a given contaminant concentration to be reached at a given depth can be used as a criterion (or determining parameter) for the groundwater vulnerability or protectability assessment. From a modeling approach, assessment by this criterion can be detailed by an analysis of the value and variation in time of the predicted contaminant concentration at characteristic depth z^* of the assessed aquifer. It is noted [*Zwahlen*, 2004] that from the viewpoint of groundwater customers (municipal services, farms, individual users) the groundwater vulnerability assessment should provide answers to three questions:

1. How long will it take for the contaminant to reach the groundwater in the case of activation of one or more surface contaminant sources?
2. What contaminant concentration will be reached in the groundwater used, for example, in the groundwater pumping wells?
3. How long contamination above the maximum contaminant level will last?

The parametric assessment by the *concentration attainment time* or the more conservative assessment by the *water percolation time* necessary for infiltrating water to reach the assessed aquifer from the surface answers only the first of these three questions, whereas the modeling assessment allows the accurate prediction of the characteristic value or even variation in time (during a definite period of interest) of the contaminant concentration at a depth of the assessed aquifer or in the water intake location.

In the a priori assessment of groundwater vulnerability, it is not known in advance what initial concentration will occur in the potential contamination source (at the day surface) or only some approximate values of the concentration can be supposed. For this reason, in the modeling assessments the relative dimensionless contaminant concentration is often considered, determined by the ratio of the assessed concentration at the given location to the source or initial source contaminant concentration. In our 1D modeling of vertical downward contaminant transport from the surface, we used relative dimensionless concentration $c(z,t)$ determined by the ratio of the dimensional predicted concentration at the assessed depth to initial surface contaminant concentration (in the infiltrating water) C_0: $c(z,t) = C(z,t)/C_0$.

The predicted value of this relative concentration for the characteristic prediction time t^* at a given depth z^* in the aquifer $c(z^*,t^*)$ characterizes the *migration permeability* of covering deposits, which can in turn serve as a measure of the *cover vulnerability* of the given aquifer to surface contamination, that is, aquifer vulnerability related to the protective role of its overlying (covering) bed of rocks and deposits.

The inverse (reciprocal) relative concentration $c^{-1}(z^*,t^*) = C_0/C(z^*,t^*)$ may serve as an indicator of protection ability (or protection potential) of covering deposits and a measure of groundwater protectability (with respect to surface contamination) of the given aquifer.

For mapping purposes, a more convenient index of groundwater protectability is the decimal logarithm of the inverse relative concentration,

$$\varepsilon = \log c^{-1}(z^*, t^*) = -\log c(z^*, t^*),$$

because values of relative concentration $c(z,t)$, especially for deep aquifers, are often small (10^{-4} and lower). A zero groundwater protectability index ($\varepsilon = 0$) corresponds to the relative concentration $c = 1$, that is, initial contaminant concentration in its source. For example, in the 1D groundwater protectability model with surface contamination, $c = 1$ at $z = 0$. The value of the groundwater protectability index $\varepsilon = 1$ corresponds to relative concentration $c = 10^{-1}$, $\varepsilon = 2$ to $c = 10^{-2}$, etc. The range of variation of the index e is determined for the whole assessed area, and starting from this range, the gradations (intervals of e value) are then chosen and subscribed to definite subareas in the course of groundwater protectability mapping. High values of ε (10 and higher) correspond to relative contaminant concentration $c = 10^{-10}$ and lower, which is most common in hydraulic and physicochemical conditions of cover groundwater protectability of a deep confined aquifer.

One should note that the introduced indicators [relative contaminant concentration in groundwater, $c(z^*, t^*)$, and groundwater protectability index ε], if taken at a depth of the groundwater table (for the upper aquifer) or upper boundary of a confined aquifer, characterize the *cover* vulnerability and protectability. Strictly speaking, this characteristic refers not to the groundwater itself but to the whole aquifer together with water-bearing rocks. If we want to consider the vulnerability or protectability of the groundwater of an aquifer in the sense of groundwater quality, then we must distinguish also that part of potential contamination that comes to deposits and rocks. For this purpose, at least the data of aquifer thickness, porosity of water-bearing formation, distribution coefficient for a particular contaminant and rock mineral composition, and hydrogeological conditions should be taken into account. This can be achieved by including the whole aquifer vertical profile with all these parameters in the assessment model. The groundwater vulnerability indicators are then calculated at some depth z^* within the aquifer, for example, in the central point of its vertical cross section or at the depth of a planned water intake well. The characteristic indicator value can also be assessed as an average for a given depth interval within the assessed aquifer or its part.

A more exact characteristic of the covering deposits (and consequently of cover's groundwater vulnerability) is the product of predicted contaminant concentration $C(z^*, t^*)$ by the downward net infiltration or groundwater recharge $w(z^*)$: $P = C(z^*, t^*) \cdot w(z^*)$, which can be called the predicted groundwater *contamination potential* at the depth z^*. This parameter, unlike the concentration itself, reflects both capacitive and screening properties of the overlying covering deposits. It is a flow-related characteristic of groundwater vulnerability. For example, if a layer of solid fractured rock with relatively thin fractures occurs on the contaminant migration pathway, the contaminant front with a relatively high concentration may rapidly reach the underlying aquifer by separate fractures, and theoretically its concentration can be measured in separate points near the fracture outlets. However, the whole average infiltration velocity w (or the downward flow rate) below the rock layer remains low because of the very thin fracture network and corresponding low layer vertical hydraulic conductivity. Thus the layer screening ability and its related protection potential remain high, which corresponds to a low contamination potential P.

Relative mapping of the predicted contamination potential for an aquifer over the given area can be done by relative contaminant concentration, $p = c(z*,t*) \cdot w(z*)$.

In our case of a 1D vertical model for the groundwater vulnerability assessment, the vertical contaminant concentration in groundwater, $C(z,t)$, is obtained from the solution of the initial-boundary problem for the advection-dispersion partial differential equation representing the 1D case of equation (1.14) [*Gladkiy et al.*, 1981]:

$$\frac{\partial}{\partial z}\left[D(z)\frac{\partial C}{\partial z} - wC \right] - \lambda C = n\frac{\partial C}{\partial t}, \tag{4.1}$$

$$C(0,t) = C_0 e^{-\lambda t}, \quad \frac{\partial C}{\partial z}\bigg|_{z=L} = 0, \tag{4.2}$$

where t is time, C_0 is initial concentration at the surface (in the infiltrating water), w (m/day) is infiltration velocity, $D(z)$ (m²/day) is the dispersion coefficient, $\lambda = \ln 2/T$ is the contaminant decay constant (T contaminant half-life period), n is the dimensionless storage coefficient,

$$n = \theta + k_d, \tag{4.3}$$

accounting for porosity or moisture content (in the unsaturated zone) θ and dimensionless distribution coefficient k_d that relates the contaminant concentrations (in Bq/dm³) in the liquid phase C and solid phase N, respectively:

$$N = k_d \cdot C. \tag{4.4}$$

Equation (4.4) represents the linear isotherm (Henry's law) for an equilibrium sorption process that is applicable for our case of relatively low contaminant concentrations in the porous media [*Bochever and Oradovskaya*, 1972; *Shestopalov*, 2002].

The approximate solution of the initial-boundary problem (4.1), (4.2) can be obtained numerically by the method of finite differences [*Gladkiy et al.*, 1981]. The problems are solved for the corresponding typical regions of the study area selected in the course of the preliminary area zoning. These regions should be characterized as well as possible by homogeneity in the area conditions of vertical flow and transport within the depth interval considered for the geological medium (soil, unsaturated zone, upper groundwater aquifer, underlying low-permeable layer, confined aquifer). All available data for the lithological composition of rocks and deposits, unsaturated zone thickness, relief, infiltration conditions, and aquifer and confining bed topology and conductivity should be taken into account. The flow and transport parameters for typical vertical boundary problems (corresponding to the selected typical regions) are determined by all these available data, including observational data from the experimental plots, literature data, and the parametric information obtained in the course of the inverse problem solutions at typical sites of the study area, which are characterized by different degrees of PFMZ occurrence and activity. The inverse problems for typical regions are most frequently solved for determination of the dispersion coefficient distribution $D(z)$ by analyzing the series of direct problem solutions for the corresponding typical region problem (4.1), (4.2) and choosing the best one (and its dispersion coefficient distribution) that gives the concentrations $C(z)$ closest to the real measured concentrations at different depths of the assessed interval [*Shestopalov et al.*, 2006].

4.2. Vulnerability and Protectability Assessment for Upper Groundwater (Unconfined Aquifer)

Assessment of vulnerability and protectability of the upper groundwater (unconfined aquifer) includes the following three stages [*Shestopalov et al.*, 2006].
1. Zoning of the study area to typical "background" regions based on the analysis of maps for unsaturated zone thickness (or depth of the upper groundwater table), infiltration conditions, relief, and the preliminary data for occurrence, character, and activity of PFMZs (depressions, lineaments) of the upper geological medium.
2. For each chosen typical region the "background" vulnerability or protectability assessment is performed for the upper groundwater aquifer (still without accounting for PFMZs) by the determination of a vertical concentration distribution $C(z,t)$, according to following procedures:
    2a. The average (by the typical region) migration parameters are determined for the corresponding initial-boundary problem of the type (4.1), (4.2), including the dispersion coefficient D, according to available literature and other data and in the course of solving the inverse problem using the observational data for concentrations at the experimental test sites.

2b. The initial-boundary problem for the typical region (4.1), (4.2) is solved numerically and the concentration distribution $C(z,t^*)$ is determined for the characteristic forecast time t^*. The forecast time t^* is determined according to the lifetime of the contaminant or from the other considerations. For example, in our case of the Chernobyl-born ^{137}Cs we considered the forecast time equal to the radionuclide half-life (30 years). In the result of the problem solution, the vertical distribution plot for the dimensionless background relative concentration $c_b(z,t^*) = C_b(z,t^*)/C(0,t^*)$ is built.

2c. For each elementary fragment of the constructed map (determined by the map detail or resolution), the depth of the groundwater table, z_1^*, is determined according to the initial data map (or the map of unsaturated zone thickness), and the value $c_b(z_1^*,t^*)$ is determined by taking the corresponding value (at depth z_1^*) from the plot $c_b(z,t^*)$ built in step 2b. This value is a measure of migration permeability of the overlying cover deposits.

Thus, from the results of steps 2a, 2b, and 2c, the map of the "background" predicted relative concentration $c_b(z_1^*,t^*)$ at the groundwater table depth is built.

3. Within each background region (according to the performed zoning at the stage 1), the procedure of refinement of the obtained "background" map of relative concentration $c_b(z_1^*,t^*)$ is performed in order to account for the typical PFMZ occurring within the study area (for example, depressions and lineaments), including the following steps:

3a. Within each background region, a more detailed (using more detailed-scale maps) zoning of PFMZs (depressions, lineaments) is performed by contouring their areas and the calculation of the fraction of their area in the total area of the typical region.

3b. The vertical predicted relative concentration profiles are calculated by the separately "adjusted" vertical models characteristic for the typical PFMZ of the given background region, according to a procedure similar to the step 2b for the background area. Depending on the required detail level and data availability, the number of PFMZ types considered (and consequently the characteristic vertical models) can be 1 (depressions), 2 (depressions and lineaments), etc. As a result, the typical vertical plots of "anomalous" relative concentration distributions $c_a(z,t^*)$ are built for each typical PFMZ of the given region.

3c. Later, similar to step 2c for the background area, the distribution of anomalous relative concentration $c_a(z_1^*,t^*)$ is found at a depth of the groundwater table within the characteristic PFMZs (depressions, lineaments, etc.).

3d. For each elementary fragment of the constructed map, the "refined" relative concentration $c(z_1^*,t^*)$ at the depth of the groundwater table, z_1^*, is determined in the following way. If the fragment belongs to the pure background zone (no depressions, lineaments, etc.), then $c(z_1^*,t^*) = c_b(z_1^*,t^*)$. If it belongs to an anomalous area of a PFMZ (of a given assessed type at step 3c), then $c(z_1^*,t^*) = c_a(z_1^*,t^*)$. If it occupies a part of background area F_b and a part of anomalous area F_a, then $c(z_1^*,t^*)$ is determined by previously assessed $c_b(z_1^*,t^*)$, $c_a(z_1^*,t^*)$ and known corresponding values of infiltration recharge at background w_b and anomalous w_a areas, according to the flow rate averaging formula:

$$c(z_1^*,t^*) = \frac{c_a(z_1^*,t^*)w_a F_a + c_b(z_1^*,t^*)w_b F_b}{w_a F_a + w_b F_b}. \tag{4.5}$$

After the above three stages, the relative contaminant concentration is assessed at a depth of the groundwater table as an indicator of vulnerability of the upper groundwater to the described contaminant.

Formula (4.5) is given for the case when the elementary fragment consists of a part of one-type PFMZ area and a part of a background area. When it consists of several parts of different-type areas (maybe several types of anomalous zones with or without the background part), the right-side part will be similar to the corresponding products in the numerator and denominator. One should note that equation (4.5) is based on the assumptions of instantaneous mixing of infiltrating waters from PFMZs and background parts of an assessed fragment of the area. In reality, during a definite time the contaminant concentration at the groundwater table depth remains higher under the local PFMZ areas (depressions, lineaments). However, the mass sum of the contaminant infiltrating into the upper aquifer in a unit of time within the assessed fragment corresponds to the average concentration determined by equation (4.5).

Values of the infiltration recharge of the upper groundwater aquifer in the background and PFMZ areas, w_b and w_a, are taken according to direct observation data obtained by lysimetric measurements at experimental test sites or balance calculations. Contouring of the PFMZ areas is performed using the topographic maps or aerospace images of higher resolution than those used for zoning of typical background areas.

In our assessment of groundwater vulnerability and protectability for the Kyiv region, the background zoning was performed using maps in the scale 1:100,000 and PFMZ zoning by maps in the scale 1:50,000 with refinement by detailed maps of key experimental sites in the scale 1:10,000.

All the above procedures of final data processing and mapping are performed either manually (which is very laborious) or with the use of GIS software, for example, ArcGIS [*McCoy*, 2004].

The obtained area distribution of dimensionless relative concentration $c_1(z_1^*,t^*)$ at a depth of the upper groundwater table is a measure of migration permeability of the unsaturated zone rocks and deposits or cover vulnerability of the upper aquifer. For an assessment of the cover protectability of the upper aquifer, the reciprocal of this value and its decimal logarithm is found:

$$\varepsilon_1 = \log c^{-1}(z_1{}^*,t^*) = -\log c(z_1{}^*,t^*), \tag{4.6}$$

This index is mapped on the final map as an indicator of cover protectability of the upper groundwater aquifer obtained taking into account PFMZs.

To assess not the cover but the *full* protectability of the (upper) groundwater that can be subjected to any risks of contamination, it is possible to consider this indicator at the higher characteristic depth $z_1{}^*$ corresponding to the real depth of the aquifer horizon from which the groundwater is pumped out: in the middle part of the aquifer taking into account all necessary model parameters of its water-bearing deposits (conductivity, porosity, distribution coefficient, etc.).

If we have real or predicted data for the initial concentration of the pore solution near the contaminated surface, then using the map of groundwater table depths (or unsaturated zone thicknesses) and the modeling vertical distributions of the dimensionless concentration for the typical areas, it is possible to obtain an approximate assessment of the predicted dimensional contaminant concentration at the groundwater table depth (for example, of ^{137}Cs in Bq/dm^3). Its map will represent the situational assessment map of cover vulnerability of the upper aquifer for the surface contaminant considered.

The obtained result can be improved by an additional expert assessment of the lithological heterogeneity of the soil and deposits of the unsaturated zone or other experimental, observation, or literature data if such data were not taken into account in the models for the typical areas. Such an improvement can be realized by introducing the recalculation coefficients for the definite zones. For example, in the case of alternating horizontal layers of sands and clays in the unsaturated zone, the vertical downward infiltration velocity is determined by the hydraulic conductivity of clays. For this reason, if such layers are present at some sites within the assessed area (in contrast to the average typical section), the assessment of groundwater protectability can be respectively improved for these sites.

The assessments obtained in the way considered above, from our viewpoint, more exactly reflect the conditions of groundwater protectability or vulnerability of the studied area with respect to the given potential or real contaminant.

4.3. Vulnerability and Protectability Assessment for Confined Aquifers

Starting from our experience in the solution of regional problems, we consider that a groundwater vulnerability and protectability assessment, especially for the confined aquifers, must be a construction element of regional works for prediction and assessment of operational groundwater resources.

From the viewpoint of modeling, the groundwater vulnerability and protectability assessments for the confined aquifers are less difficult tasks as compared to the upper groundwater. This is caused by less complicated and easier predictable physical processes of groundwater flow and migration in the saturated medium, as compared to the moisture transport in the unsaturated zone.

There are a lot of well-developed modeling methods and computer codes for the prediction of groundwater flow and transport in the stored groundwater aquifers. Among them the most known are MODFLOW and MT3D [*McDonald and Harbaugh*, 1988; *Zheng*, 1990; *Ciang and Kinzelbach*, 2001].

On the other hand, because of the higher depth of the confined groundwater, the lack of data for lithology and permeability of water-bearing deposits at depths of confined aquifers can be more significant than for the upper groundwater. In this relation, significant variations exist in the assessments of flow and transport parameters of geological media such as hydraulic conductivity, dispersion, and distribution coefficients. In conditions of insufficient degrees of study of the flow and transport parameters for the confined aquifers, it is possible to apply for their vulnerability and protectability assessment a simplified methodology similar to the one described above for the unconfined upper groundwater. The methodology is based on combining the zoning methods and 1D modeling of the contaminant's downward migration from the previously assessed upper aquifer (see Section 4.1) into the given confined aquifer through the low-permeable confining bed. The corresponding 1D model is described by the same general formulation (4.1)–(4.4).

As the upper model boundary for the confined aquifer model, the characteristic depth z_1^* of the previously performed assessment of the upper aquifer is taken. The model is formulated in terms of the relative dimensionless concentration in such a way that the upper boundary relative concentration value at depth z_1^* is taken to be 1. In such a way the protection ability of the whole overlying deposit from the contaminated surface, including the previously assessed upper aquifer, is taken into account.

The vertical downward flow velocity in the confining bed between the first and second (confined) aquifers, w (m/day), is determined in accordance with assumptions of the vertical flow between the aquifers by the formula [*Polubarinova-Kochina*, 1977; *Ciang and Kinzelbach*, 2001]

$$w = \frac{k_0}{m_0}(H_1 - H_2), \qquad (4.7)$$

where k_0 (m/day) and m_0 (m) are vertical hydraulic conductivity and thickness of the confining bed, respectively, and H_1 and H_2 are groundwater heads in the first (unconfined) and second (confined) aquifers, respectively. These parameters are determined by available literature data, observations in the monitoring wells, and results of modeling.

Of primary importance is the sign of the difference $\Delta H = H_1 - H_2$. At positive ΔH the flow across the confining bed is directed downward, and at negative ΔH it is directed upward. In the latter case the uprising flow prevents the downward migration of contaminants from the upper aquifer into the assessed confined aquifer. At such sites the confined aquifer (in the case of a well-occurring confining bed) should be regarded as conventionally protected.

The main transport parameters are as follows:

Storage coefficient n determined by equation (4.3)

Dispersion coefficient D (m²/day) that in the given case of saturated medium can be related with vertical flow velocity w by the formula

$$D = \alpha \cdot w, \tag{4.8}$$

where α *(m)* is the longitudinal dispersivity coefficient [*Bochever and Oradovskaya*, 1972; *Ciang and Kinzelbach*, 2001].

After determination of the initial parameters, the corresponding boundary problem of the type (4.1)–(4.4) formulated for the relative concentration $c_2(z,t)$ of the confining bed and second confined aquifer is solved numerically for the characteristic prediction time t^*, and the obtained vertical concentration distribution $c_2(z,t^*)$ is plotted against depth z. Usually, for the study area the vertical flow velocity assessed by equation (4.7) varies within a definite range. Taking the series of possible values for the velocity w (with a definite step Δw), a series of resulting plots for corresponding vertical concentration distributions can be built. The vertical coordinate z in the model is local, with zero mark $z = 0$ corresponding to the absolute depth z_1^* of the previous assessment for the upper unconfined aquifer, and the obtained vertical plots for the concentration $c_2(z,t^*)$ are in the absolute depth range between z_1^* and z_2^*, where the latter corresponds to the characteristic assessment depth of the assessed confined aquifer.

Further vulnerability and protectability assessment procedures for the confined groundwater aquifer are as follows:

1. The study area is ranged into area zones with approximately equal thicknesses of the confining bed m_0 (map 1) with required detail for the observed range of m_0 variation (for example, by average thicknesses 10, 20, 30, and 40 m).

2. In the same way, the study area is then ranged into area zones with approximately equal vertical flow velocity w (map 2) using formula (4.7). The detail of these zones depends on the available parametric information for k_0, m_0, H_1, and H_2. The independent groundwater flow modeling, including the 3D hydrogeological model, is very helpful at this stage.

3. If the data are available for deep PFMZ occurrences at the depth's level of the confining bed, then they must be taken into account by corresponding zones of relatively low confining bed thickness and/or higher vertical flow velocity. Sites of the confining bed thinning-out or its breakthrough zones (often called "hydrogeological windows") are contoured, especially those where ΔH is

positive. Such sites require special attention with respect to the corresponding vertical flow velocity w assessment because in cases of m_0 being close to zero, equation (4.7) may lead to high inaccuracy. However, in the real situation of a "hydrogeological window," instead of low-permeable deposits of the confining bed as in the background undisturbed area, there are higher-permeable deposits (for example, sands or loamy sands instead of loams) in such a window. For this reason, equation (4.7) can still be formally used (under higher attention), and there is no need to set m_0 to zero, but it is more correct to increase (usually 5–10 times or even more) the vertical conductivity k_0. If deep PFMZs related to the discontinuity of the confining bed widely occur and occupy a significant part of the total study area, they require a separate typical model to be implemented and a corresponding vertical predicted concentration profile $c_2(z,t^*)$ obtained for these zones.

4. Next, the obtained maps for zones of the confining bed thickness (map 1) and vertical flow velocity (map 2) are overlaid and the resultant zoning map is obtained. Each zone on this map corresponds to definite average values of the confining bed thickness m_0 and vertical flow velocity w.

5. For each obtained zone with definite average values of m_0 and w, the absolute depth mark z_2^* is found on the vertical axis Z, corresponding to the assessment depth of the confined aquifer. If the assessment depth z_2^* corresponds to the upper boundary of the confined aquifer (just below the confining bed), then the assessment gives the *cover* vulnerability or protectability of the confined aquifer. For the assessment of full groundwater vulnerability or protectability, it is necessary to relate z_2^* with a depth of the real groundwater intake (within the aquifer body). Then the corresponding local model coordinate $z = z_2^* - z_1^*$ is found, and by the intersection of this depth level with the corresponding plot of predicted relative concentration $c_2(z,t^*)$ for the given vertical flow velocity w, the relative concentration at the absolute depth of the assessed confined aquifer, $c_2(z_2^*,t^*)$, is determined. This relative concentration characterizes the additional vulnerability of the confined aquifer determined by the state of the permeability of the overlying confining bed. The area distribution is then calculated for the additional groundwater protectability index $\varepsilon_2 = -\log c_2(z_2^*,t^*)$ on account of the attenuation capacity of the confining bed. The working map for this index is then built.

6. The final assessment is performed of the cover or full groundwater protectability of the second confined aquifer, from the surface accounting for the previously assessed protectability of the upper aquifer (see Section 4.1), by overlaying the working maps and summing the corresponding indexes ε_1 and ε_2 of the first and second aquifers for the corresponding characteristic vertical profiles:

$$\varepsilon = \varepsilon_1 + \varepsilon_2 = -\log c_1(z_1^*,t^*) - \log c_2(z_2^*,t^*). \qquad (4.9)$$

For the overall confined aquifer vulnerability assessment, the working maps for relative concentrations for the first and second aquifers are overlaid, and the product of the corresponding assessed relative concentrations is calculated for each vertical characteristic profile corresponding to the mapping fragment:

$$c(z_2{}^*,t^*) = c_1(z_1{}^*,t^*) \cdot c_2(z_2{}^*,t^*). \tag{4.10}$$

The obtained map for the resultant relative concentration $c(z_2^*,t^*)$ characterizes the vulnerability of the second confined aquifer to the surface contamination.

In the same way as for the upper aquifer, the *predicted contamination potential* of the confined aquifer, $p = c(z_2^*,t^*) \cdot w$, can be calculated and may serve as a more refined flow-related characteristic of confined groundwater vulnerability.

For an assessment of the full groundwater vulnerability of the confined aquifer, it is necessary to take the assessment depth z_2^* in the bulk of the aquifer, for example, in its middle point by depth. On the modeling profile, in the course of model implementation for the characteristic typical zones of the area, the depth interval belonging to the aquifer bed is characterized by its corresponding flow and transport parameters, especially by the higher hydraulic conductivity k_2 (as compared to k_0 of the overlying confining bed). In this case, when using equation (4.7) for the calculation of the vertical flow velocity w, instead of k_0, the average hydraulic conductivity value (between the confining bed and the aquifer bed) k_{02} should be used and can be determined by equation (1.5) (see Chapter 1). This note is also significant for the upper aquifer; that is, the average vertical flow velocity w should be assessed taking into account the vertical hydraulic resistance in the depth interval between the assessment depths of the first and second aquifers, z_1^* and z_2^* (Figure 4.1).

The described methodology of groundwater vulnerability and protectability assessment for the second (confined) aquifer from the surface can also be applied for even deeper aquifers (third or fourth from the surface and deeper). In this case, in the same way as for the second aquifer, at the preparatory stage of analysis of the groundwater heads (levels), the zones of downward (mostly within watersheds) and upward (in the river valleys) flow are found as shown in Figure 4.1 [*Shestopalov*, 1988].

Starting from the upper aquifer, assessment is performed for the zones of downward flow successively for each aquifer at control depths z_1^*, z_2^*, z_n^* according to the procedure described above.

As a result, in the same way as in equation (4.9) for two aquifers, for the aquifer system of n aquifers we obtain the area distribution of the cover or full groundwater protectability ε depending on the assessment depth z_n^* determined by the sum of n partial indexes for the successive aquifers of the system:

$$\varepsilon = -\log c(z_n{}^*,t^*) = -\sum_{i=1}^{n} \log c_i(z_i{}^*,t^*), \tag{4.11}$$

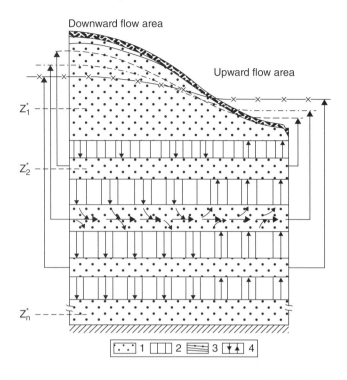

Figure 4.1 Flow scheme of a stored aquifer system [*Shestopalov*, 1988]: (1) aquifers; (2) low-permeable beds; (3) groundwater heads of stored aquifers; (4) groundwater flow directions.

In the same way as for equation (4.10), the resulting relative concentration of the aquifer number n, $c(z_n{}^*, t^*)$, is determined by the product of separate partial relative concentrations for each aquifer number i:

$$c(z_n{}^*, t^*) = \prod_{i=1}^{n} c_i(z_i{}^*, t^*) \tag{4.12}$$

It can be considered and mapped as the indicator of vulnerability of the aquifer number n to surface contamination.

We note once again that to refine the cover vulnerability assessment of the aquifer to a full assessment of the aquifer's groundwater, we must account also for the additional sum of physical and geochemical barrier properties of the aquifer represented by the hydraulic resistance, porosity, and sorption capacity of the water-bearing rocks and deposits of the aquifer. In the course of the corresponding modeling of the vertical contaminant transport, the following are required:

1. An initial setting of (as accurate as possible) flow and transport parameters such as hydraulic conductivity, porosity, distribution coefficient b [equation (4.3)], or K_d [see equation (1.9), Chapter 1] over the whole thickness of the aquifer

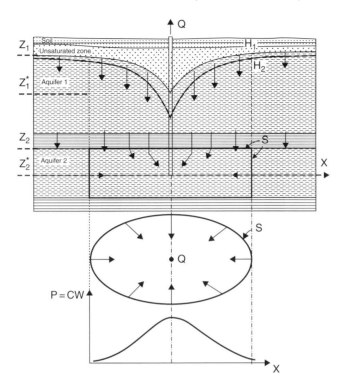

Figure 4.2 Scheme of groundwater vulnerability assessment for a separate water intake well.

2.   Consideration of characteristic depth z_n^* of the assessed groundwater in the aquifer body taking into account the whole vertical profile on the contaminant's pathway from the surface

The contamination potential of the aquifer number n from the surface can be determined by the formula

$$p_n = c(z_n^*, t^*) w_n. \qquad (4.13)$$

One can note that the considered method characterizes the groundwater vulnerability and protectability of the upper aquifer and deeper confined aquifers from the viewpoint of downward vertical contaminant transport. Such an assessment is sometimes called the *resource* groundwater vulnerability assessment [*Zwahlen*, 2004], in contrast to the *targeted* assessment for a separate water supply source (well, water intake, etc.). In the latter case, with the vertical flow and transport accounting for disturbed flow conditions such as increased vertical head gradients around the well or water intake, one should also consider lateral groundwater flow and transport in the aquifer to the well within its recharge area, as is shown in Figure 4.2. The predicted concentration in the pumped groundwater will be determined by the average contamination potentials of the downward infiltration flow $p_z = c(z_2)w_z$ and lateral flow $p_x = c(z_2^*)w_x$ (possibly of cleaner water with lesser contamination potential).

In the 3D case of the well in Figure 4.2, the average contaminant concentration in the pumped groundwater can be represented in the form

$$C_Q = \frac{\iint\limits_S P_n \, ds}{\iint\limits_S W_n \, ds},$$ (4.14)

where W_n and P_n are components of the flow velocity and contamination potential normal to the integration surface S bounding the recharge area of the well from the above and lateral sides. It is clear that for characterization of groundwater vulnerability in this case the 1D modeling approach is already insufficient, and a complete 3D model described by flow and transport equations (1.14)–(1.17) (see Chapter 1) should be considered. This 3D model can be based, for example, on the codes MODFLOW and MT3D [*Ciang and Kinzelbach*, 2001].

In the simplified 1D model of groundwater vulnerability and protectability assessments for stored aquifers, the PFMZs correspond to zones and sites of increased vertical flow velocity w as compared with corresponding background areas.

On the whole, one should note that clarifying the real input of the anomalous flow and transport paths (or PFMZs) in the contamination of the subsurface hydrosphere is a very complicated problem. It cannot be solved based only on the general qualitative consideration of the complicated hierarchy system of preferential flow pathways or separate elements of this system. Only the implementation of special observations and experimental studies at pilot plant sites allows the solution of this problem taking into account the entire effect of PFMZs and preferential flow pathways on the groundwater contamination.

5. GROUNDWATER VULNERABILITY AND PROTECTABILITY TO CHERNOBYL-BORN RADIONUCLIDE

5.1. Upper Groundwater

Let us consider the methodology described in Chapter 4 assessment of groundwater vulnerability and protectability as related to contamination by Chernobyl-born radionuclide ^{137}Cs in the study area of the Dnieper River basin within the borders of the Kyiv region.

According to the developed methodology (see Chapter 4), we have performed preliminary zoning of the study area into three landscape types: (1) the southern areas of black soils and loess-like loams; (2) the central (Kyiv) areas of loamy and sod-podzolic soils and sandy-loamy composition of the unsaturated zone; and the (3) northwestern areas of Polesye, the first and second Dnieper floodplain terraces and sandur plains.

In accordance with the methodology described in Chapter 4, for each of these typical landscapes we obtained representative vertical distributions of the relative concentration of the radionuclide to a depth of 20 m (upper part of the Quaternary aquifer) for the 30 year period (the half-life for ^{137}Cs). In the majority of the area, this period corresponded to the achievement of real measurable concentrations of the radionuclide in the groundwater. The typical vertical distributions of the predicted relative concentration were calculated using the vertical transport model (4.1)–(4.4) after the corresponding model calibration for each of three characteristic landscape types described above.

During the model calibration, the infiltration velocity w and storage coefficient $n = \theta + k_d$ [see equation (4.3)] were determined using the available observation and experimental data for the sites within the characteristic landscape areas [*Shestopalov*, 2001]. The values of w varied for different landscape types: from 50 mm/year (landscape type 1) to 75 mm/year (landscape type 2) to 100 mm/year (landscape type 3). The values of n are determined mainly by the distribution coefficient k_d characteristic for the given radionuclide (^{137}Cs) because it is usually several orders of magnitude higher than the effective porosity or moisture content, θ,

Groundwater Vulnerability: Chernobyl Nuclear Disaster, Monograph Number 69.
Edited by Boris Faybishenko and Thomas Nicholson.
© 2015 American Geophysical Union. Published 2015 by John Wiley & Sons, Inc.

which ranges from 0.1 to 0.5. The value of k_d is determined as a ratio of radionu-clide concentrations in groundwater (see Table 2.1 and Figure 2.1) and solid phase (see Figures 2.3, 2.4, and 2.5) recalculated to volume concentrations. It depends on the soil and rock types reaching its maximum in the upper soil and decreasing with depth usually by 1–2 orders of magnitude. According to our previous assess-ments [*Shestopalov*, 2001], its value was taken as 10–100 in landscape type 1, 5–50 in landscape type 2, and 1–10 in landscape type 3. The vertical distributions of the dispersion coefficient $D(z)$ for three typical sections were determined during the inverse problem solution for the calibration period 10 years (year 1996) for which the data of groundwater sampling were available. The obtained values of D vary from 0.001 m²/day (landscape type 1) to 0.01 m²/day (landscape type 3).

It is supposed that the dispersion coefficient obtained by this calibration procedure should to some extent implicitly account for the flow and transport heterogeneities of the geological medium.

Figure 5.1 gives an example of the model calibration plot of relative ^{137}Cs concentration against depth for a typical depression area in the CEZ (landscape type 3) with its corresponding sampling data points obtained in 1996.

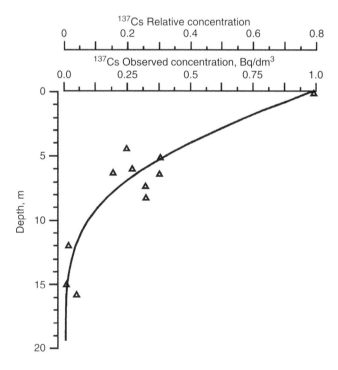

Figure 5.1 Model calibration plot of relative ^{137}Cs concentration against depth (forecast time 10 years) for landscape type 3 with corresponding sampling data points obtained from wells and soil pore solution (upper point) in a typical depression within CEZ in 1996.

The two horizontal axes on the plot show the observed concentrations in Bq/dm^3 (lower axis) and corresponding dimensionless concentrations relative to initial values in the soil solution (upper axis).

After determination of the initial flow transport parameters during the model calibration for a 10 year time period, the forecast vertical distributions of the relative ^{137}Cs concentration for a 30 year period were obtained. The forecast (30 year) and calibration (10 year) relative concentration plots against depth for the above three representative types of areas are shown in Figure 5.2 (plots 1, 2,

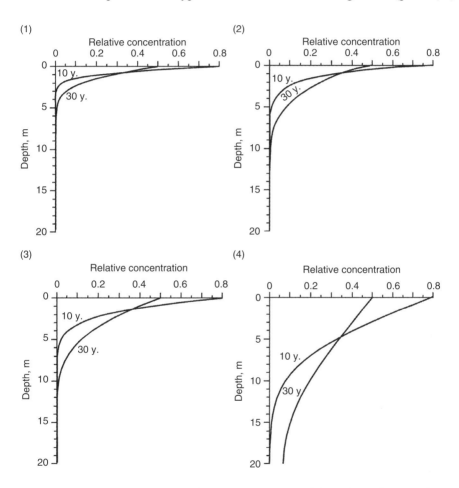

Figure 5.2 Calibration (10 year) and forecast (30 year) modeling plots of ^{137}Cs relative concentration in groundwater (percent of initial surface concentration in liquid phase) against depth at typical areas of the Dnieper basin: (1) territories of black soils and loess-like loams; (2) territories of sod-podzolic soils and loamy-sandy compositions of the unsaturated zone; (3) territories of floodplain terraces and sandur plains; (4) typical depression of floodplain terraces and sandur plains.

and 3, respectively). As is seen, the predicted concentration values at definite depths increase successively with changes in the characteristic landscape type from 1 to 3. The increase is not prominent when changing from plot 2 to plot 3. This can be explained by the relatively similar sandy-loamy composition structure of the aeration zone in these two subregions. However, in correlation with the surface contamination density, the dimensional ^{137}Cs concentrations in Bq/dm^3 in the upper groundwater may increase 1–2 orders of magnitude with changes in the landscape type from 1 to 2 and from 2 to 3. To obtain refined groundwater vulnerability and protectability maps accounting for depression-related PFMZs, a similar modeling was performed using data for typical depressions within the CEZ and Kyiv regions. The corresponding vertical profile of the dimensionless concentration in floodplain areas (subregion 3) is shown in Figure 5.2, plot 4.

The calculated characteristic profiles for the above three typical areas are used to prepare the preliminary ("background") groundwater vulnerability and protectability maps for ^{137}Cs within the Dnieper basin.

Following the zoning of the study area into three typical subareas and obtaining the typical vertical concentration distributions, the assessment depths z_1* have been specified according to available data for the upper groundwater table depth and the values of the predicted relative ^{137}Cs concentrations for a 30 year period taken at these depths from the typical vertical distributions. The obtained area distribution of the relative concentration at depths of the groundwater table characterizes the background cover vulnerability of the upper Quaternary aquifer (still with no account of PFMZs).

Further on, the refinement stage accounting for depression-related PFMZs has been performed according to the typical vertical distribution of relative ^{137}Cs concentrations (curve 4, Figure 5.2).

By the available observation data for the infiltration rate w at a groundwater table depth and the obtained predicted concentration of ^{137}Cs, the approximate input of depression-related PFMZs into the total infiltration recharge and their possible share in the total contamination of the Quaternary aquifer have been determined.

To this end, a cartographic analysis of depression occurrence was performed using topographic maps on scales of 1:50,000 and 1:20,000 and separate local areas on a scale of 1:10,000. The analysis shows that the number of contoured depressions per unit area increases significantly with increasing detail of the map. However, even the less detailed scale of 1:50,000 enabled more than 2000 depressions to be discovered per standard map sheet on the scale 1:50,000. Using these data, the coefficient $K_c = S_{10}/S_{50}$ determined by the ratio of the total depression area identified by the map on a scale of 1:10,000 to that identified by the map on the scale 1:50,000 was found, and its dependence on the total number of depressions per standard map sheet on the scale 1:50,000 was plotted (Figure 5.3).

Using this plot and a sheet-by-sheet calculation of the depression area of this scale, a map of the relative area density of depressions (in percentage of a

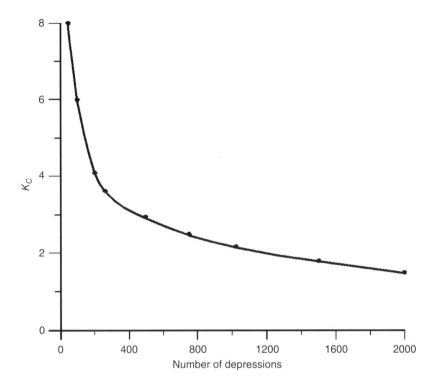

Figure 5.3 Plot for the ratio K_c of total depression area identified by the map on a scale of 10,000 to that of 1:50,000 on depression number per standard map sheet of 1:50,000.

standard map sheet area in the scale 1:50,000) for the studied territory of the Kyiv region was drawn (shown in Figure 5.4). According to this scheme, the areal share of depressions increases from 0% in the southern part to 10–15% in the eastern part of the region.

Returning to the PFMZ classification given in Chapter 3, it is worth noting that the depression-related PFMZ characteristics of plain areas described here along with the gullies and ravines characteristic of the dissected relief range in linear dimension from 10 to 10^3 m, so they can be categorized as mesozones and microzones. The PFMZs with dimensions of 10^3–10^5 m (macrozones and megazones) are taken into account in the course of preliminary area zoning on a scale of 1:50,000 and in part on a scale of 1:20,000. The PFMZs with dimensions ranging from 10^{-2} to 10 m (femtozones, picozones, and nanozones) are in most cases implicitly taken into consideration by the groundwater sampling procedure.

In the same way the assessment can be implemented for PFMZs associated with linear geodynamical zones (lineaments).

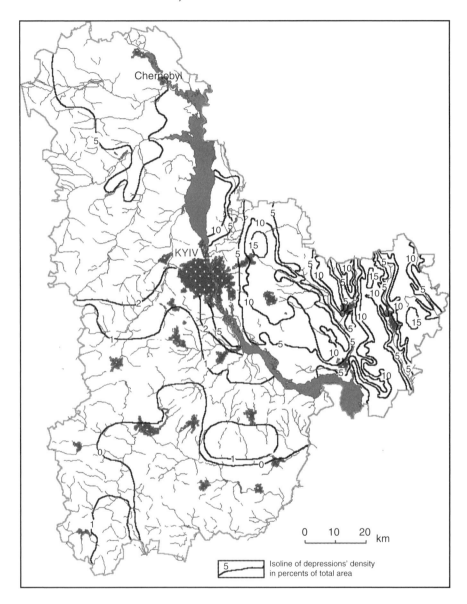

Figure 5.4 Distribution scheme of depression densities (percent of area covered by depressions) for the Dnieper basin (Kyiv region) obtained from a cartographic analysis in scale 1:50,000. (*For color detail please see color plate section.*)

After the assessment of average contaminant concentration in the ground-water taking into account PFMZs according to formula (4.5) (see Section 4.1), a map has been drawn of the cover groundwater protectability of the Quaternary aquifer with respect to ^{137}Cs within the study area. It represents the distribution in the area of the groundwater protectability index $\varepsilon_1 = -\log c(z_1^*, t^*)$, determined by equation (4.6), where $c(z,t)$ is the dimensionless relative concentration (see Figure 5.5A).

The map depicted in Figure 5.5 (Legend A) reflects the attenuation capacity of deposits in the unsaturated zone, with no reference to distribution of the soil contamination. Hence, it corresponds to the assessment of the intrinsic cover protectability of the upper Quaternary aquifer.

Further on, to obtain an approximate assessment of the groundwater occurring deeper in the middle aquifer body, a repeated procedure determining the average relative concentration of radionuclides has been performed in the same regions (gradation zones) as for the preliminary cover aquifer protectability vulnerability assessment. For this purpose, the predicted concentration values have been taken at depths z_1^*, corresponding to the middle point (in terms of thickness) of the aquifer body (see Section 4.1). For these concentration values the corresponding values of the groundwater protectability index ε_1 have been calculated and are shown in Figure 5.5B.

As seen in Figure 5.5, the radionuclide penetrates into the aquifer throughout most of the study area; however, its relative concentration is different, as is shown by the gradations of the concentration range: 0–2%, 2–5%, 5–10%, 10–20%, and over 20% of the surface concentration (in the liquid soil phase).

Overall, the Kyiv region is characterized by comparatively low values of the groundwater cover protectability index for the Quaternary aquifer. It varies from below 0.7 to 2 with corresponding relative ^{137}Cs concentrations from 20% to 1% of the soil liquid-phase concentration. The minimum values of the aquifer protectability index correspond to the northern territories of Ukrainian Polesye, valleys of the Dnieper and its tributaries (Figure 5.5). The highest aquifer protectability index $\varepsilon > 1.5$ ($c < 3\%$) corresponds to elevated, often forested areas with maximum thicknesses of the unsaturated zone or depth of the groundwater table. Characteristic for these areas are loamy and sod-podzolic soils and sandy-loamy compositions of the unsaturated zone. Within the Kyiv conurbation area, the upper aquifer protectability index varies from below 0.7 to 1 (corresponding to relative concentrations of ^{137}Cs from over 20% to 10%). It is sometimes higher at elevated areas of the Dnieper right bank (the watershed area east of the Irpen River), reaching 1.5 ($c = 3\%$).

The contaminant relative concentration in the groundwater of the middle aquifer body (Figure 5.5, Legend B) appears to be lower than at the groundwater table depth. For the corresponding gradations it varies within 0–0.1%, 0.1–1%, 1–5%, 5–10%, and over 10% of the upper soil liquid-phase concentration. Corresponding values of the groundwater protectability index ε vary from below 1 to over 3.

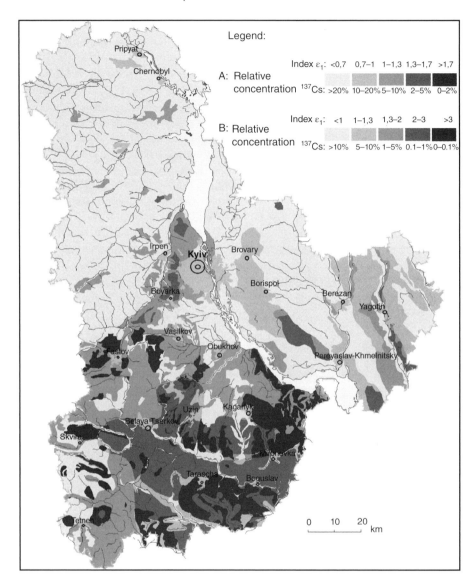

Figure 5.5 Map of cover protectability of Quaternary aquifer and upper groundwater protectability from ^{137}Cs contamination for the Kyiv region for a prediction period of 30 years: (A) index of cover groundwater protectability $\varepsilon_1 = -\log c$ (see Table 5.1) and predicted relative concentration of ^{137}Cs at a groundwater table depth (in percent of near-surface concentration in infiltrating water); (B) index of groundwater protectability and predicted relative concentration in the middle aquifer body.

The drawn map of groundwater protectability (Figure 5.5) in the inverse sense (by gradations of relative radionuclide concentration) also characterizes the intrinsic (relative) groundwater vulnerability with no regard to the real contamination source (surface ^{137}Cs contamination distribution).

To obtain the Chernobyl case-related assessment (by the real observed surface contamination with ^{137}Cs) of the upper aquifer vulnerability, the relative groundwater protectability map obtained above was superimposed ("overlaid") with the map of the postaccident soil ^{137}Cs contamination density. The values of soil contamination density taken from the corresponding map were multiplied by the corresponding values of the predicted relative concentration for corresponding gradation zones of the relative groundwater protectability map. Therefore, an assessment map of groundwater vulnerability to real surface contamination by ^{137}Cs was obtained in gradations of equivalent area contamination density (Figure 5.6).

According to the applied methodology, when obtaining the vulnerability map of the upper groundwater to real surface contamination with ^{137}Cs, it would be more correct to overlay the protectability map (Figure 5.5) with an area map of the initial radionuclide concentration in the infiltrating near-surface water. However, due to the lack of available data for radionuclide concentrations in the near-surface pore solutions, we used the conservative assessment of the mobile radionuclide forms in the total surface soil contamination.

Starting from the obtained range of total radionuclide activity incoming to the aquifer per unit area, five gradations were determined for the characterization of the aquifer vulnerability to ^{137}Cs. In this case the characteristic value of groundwater vulnerability, instead of the concentration in groundwater, is expressed in units of equivalent surface radionuclide density (kBq/m^2, column A, or Ci/km^2, column B, Figure 5.6).

Further on, using the obtained sum of contamination per unit area and average aquifer parameters for the study area including the saturated zone thickness, effective porosity $n = 0.2$, average rock density 2.6 g/cm^3, and the average distribution coefficient for ^{137}Cs, $K_d = 10$ dm^3/kg [*Shestopalov*, 2001], an attempt was made to assess (for the same specified zones) the full groundwater vulnerability through the calculation of the corresponding approximate absolute concentration of ^{137}Cs in the aquifer groundwater (in mBq/dm^3, column C, Figure 5.6). This assessment corresponds well with available sampling data for the Quaternary aquifer (see Table 2.1).

According to the obtained map (Figure 5.6), the areas of maximum groundwater vulnerability to ^{137}Cs (concentrations over 100 mBq/dm^3) correspond to the areas of highest surface radionuclide contamination in the northwestern part of the CEZ. The relatively high groundwater vulnerability values with predicted concentrations over 10 mBq/dm^3 are characteristic for the northern areas of Polesye and spots of high radioactivity in the central part north of Kyiv city.

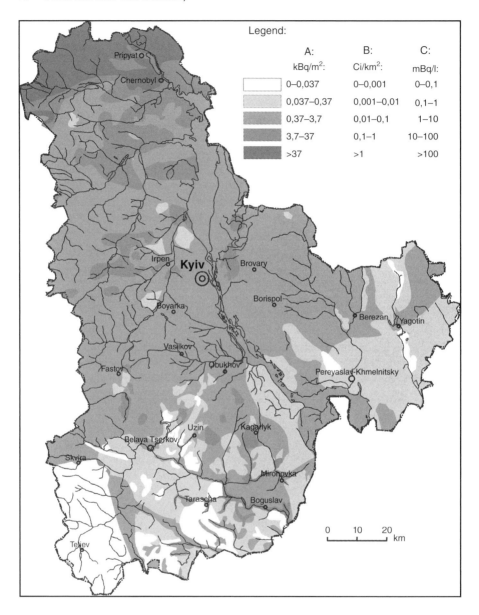

Figure 5.6 Map of predicted vulnerability of Quaternary aquifer and its groundwater within the Kyiv region to ^{137}Cs composed using a postaccident map of surface contamination density: (A, B) assessment of cover aquifer vulnerability in units of surface contamination density (kBq/m^2 and Ci/km^2); (C) assessment of full groundwater vulnerability accounting for the attenuation capacity of aquifer water-bearing rock (in units of dimensional predicted concentration, mBq/dm^3).

The minimum-vulnerability (^{137}Cs concentrations below $0.1\,\text{mBq/dm}^3$) and vulnerability groundwater areas correspond to relatively small sites of watershed areas in the south and southwestern parts, respectively, of the Kyiv region and to separate relatively clean spots. The rest of the territory, which is the largest by area, corresponds to average-vulnerability groundwater with ^{137}Cs concentrations in the range of 1–$10\,\text{mBq/dm}^3$.

To evaluate the proposed method, the results obtained were compared with the results of the assessment of groundwater vulnerability to ^{137}Cs for the same territory of the Kyiv region by the Russian method [*Belousova*, 2005], which was vulnerability described in Chapter 1. As already mentioned, we performed it in the course of the joint Russian-Ukrainian-Belarusian project in 2003 [*Shestopalov*, 2003]. A schematic map of the assessment is shown in Figure 5.7. As can be seen in Figure 5.7, the obtained vulnerability variation over the area is close to the actual contamination of the surface of the Kyiv region (see Figure 2.2). Maximum values of groundwater vulnerability (very high and high) occur in the northern part of the territory (CEZ). Conditionally invulnerable (very low) groundwater is in the central, southern, and southeastern parts of the region covering most of the study area. Low-vulnerability groundwater corresponds to individual spots of increased radioactivity in the southern part of the region.

As one would assume, the groundwater vulnerability estimated using our proposed procedure (Figure 5.6) is significantly higher over the study area than that obtained by the Russian method. It should be emphasized that in the Russian method the vulnerability categories are determined depending on the contaminant travel time from the surface to the groundwater table: up to 30 years, 30–60 years, 60–100 years, and more than 100 years (see Table 1.5). As a result, for up to 30 years the radionuclides could reach the groundwater level within only a few sites in the northern region.

However, actual data show that quite measurable radionuclide concentrations (see Table 2.1) were observed already a few years after the accident (1992–1997) not only within the CEZ area but also in the central part of the region (see Figure 2.1).

According to the proposed methodology, the obtained vulnerability map is drawn for a forecast period of 30 years from the onset of contamination. The map gradations are determined by the relative concentration and expressed in absolute units based on the value of the actual surface contamination. A separate scale (C) shows an approximate estimate of the total groundwater vulnerability based on the protective capacity of the rocks in the accessed aquifer. As can be seen in Figure 5.6, unlike Figure 5.5, within almost the entire territory of the region there is penetration of the pollutant on the groundwater table. However, the activity of the penetration is different. It not only depends on the differences in surface contamination density and the protective ability of the vadose zone but also accounts for PFMZ distribution and activity in the study area.

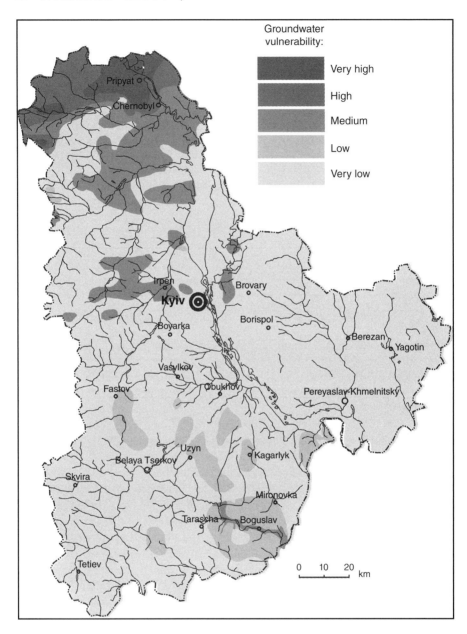

Figure 5.7 Map of groundwater vulnerability to ^{137}Cs of Quaternary aquifer for Kyiv region area [*Shestopalov*, 2003] using the Russian methodology [*Belousova and Galaktionova*, 1994].

5.2. Confined Aquifers

Using the methodology considered in Chapter 4, we performed an assessment of groundwater vulnerability and protectability to Chernobyl-born ^{137}Cs for the second (from the surface) confined Eocene aquifer (depth 60–130 m) separated from the upper Quaternary aquifer by a low-permeable bed of Kyiv marls with thicknesses of up to 50 m. The assessment was done for the study area of the Kyiv region within the Dnieper River basin.

To define the initial flow parameters, the corresponding data were taken from the 3D hydrogeological model developed previously for the Kyiv area. The model is based on the regime observations, literature data, and results of the inverse problem solution [*Rudenko et al.*, 1997; *Shestopalov et al.*, 1997]. The assessed area covered by the model comprises 22,000 km^2. The minimum dimension of the rectangular blocks of the model was 4×6 km, which determined the detail of the assessment.

Based on these data, the vertical flow velocity component w (m/day) was determined using equation (4.6). During the assessment of the vertical permeability for the low-permeable bed of Kyiv marls, k_0/m_0, values of vertical hydraulic conductivity k_0 were determined to be in the range of 10^{-4}–10^{-3} m/day, and the bed thickness m_0 ranged from 0 (in sites of thinning) to 10–20 m in river valleys and 20–40 m in the water-divide areas. To determine the vertical flow velocity w, the less significant additional hydraulic resistance component (as compared to the low-permeable bed) of the included part of the upper Quaternary aquifer (its m/k) was also taken into account, and equation (1.5) was used to determine the total vertical hydraulic conductivity (see Chapter 1).

The obtained values of the vertical flow velocity w within the study area (except the sites with upward flow) vary from 0 to 2000 mm/year. Such a wide range can be explained by the presence of operating water intakes in the Kyiv region which exploit the underlying Cenomanian-Callovian and Bajocean aquifers and cause the development of depression cones in these aquifers that lead to significant downward flow velocities in the studied Eocene aquifer.

The depth of the upper 1D model boundary described by equations (4.1)–(4.4) for the calculation of vertical relative concentration distributions corresponded to the level z_1^* of the groundwater table in the upper aquifer, for which the previous assessment of groundwater vulnerability and protectability was performed.

The transport parameters of the saturated part of the Quaternary aquifer and low-permeable Kyiv marl bed were specified by the conservative assessment based on literature data. The storage coefficient n, determined by equation (4.3), for sands and loamy sands of the upper Quaternary aquifer was taken to be 0.6 (the sum of distribution coefficient $k_d = 0.5$ and porosity $\theta = 0.1$), and for the low-permeable bed of Kyiv marls n was equal to 1.01 (including $k_d = 1$ and $\theta = 0.01$). This was a conservative assessment according to the data of *Bochever and Oradovskaya* [1972]. The dispersion coefficient D (m^2/day) was calculated by formula (4.8), in which the dispersivity coefficient α for the confining bed of Kyiv marls was taken as 0.1, in agreement with literature data for low-permeable rocks [*Bochever and Oradovskaya*, 1972].

After the initial parameters were defined, the initial-boundary problem (4.1), (4.2) was solved numerically using the method of finite differences [*Gladkiy et al.*, 1981], and the vertical distributions of relative ^{137}Cs concentrations in the groundwater were calculated for different values of downward flow velocity w within its variation range from 10 to 2000 mm/year. After that, the corresponding assessment plots of relative ^{137}Cs concentration $c_2(z,t^*)$ for the forecast period $t^* = 30$ years (the half-life time for ^{137}Cs) were drawn.

The obtained assessment relative concentration plots (as referenced to the initial concentration of 1 at a depth of the Quaternary aquifer groundwater table) are shown in Figure 5.8 against depth in the local scale with zero corresponding

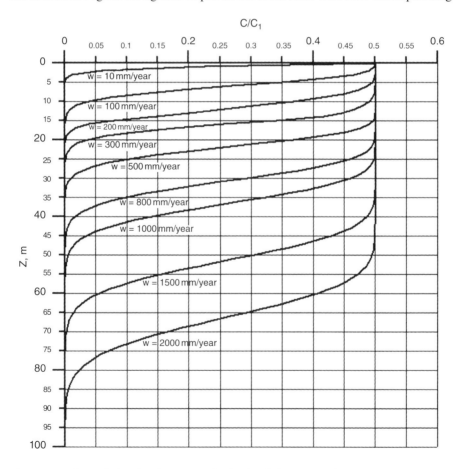

Figure 5.8 Plots of the distribution with depth of the ^{137}Cs relative concentration in groundwater infiltrating through the low-permeability bed Kyiv marls corresponding to different values of vertical flow velocity w for the forecast period $t^* = 30$ years.

to the upper aquifer assessment depth z_1^*. The plots shown in the figure correspond to the indicated values of vertical flow velocity w.

Further on, according to the procedure described in Section 4.2, maps were drawn of the low-permeability confining layer thickness m_0 (Figure 5.9) and calculated vertical flow velocity w (Figure 5.10) according to data taken from the 3D model of the Kyiv region [*Rudenko et al.*, 1997; *Shestopalov et al.*, 1997].

The thickness m_0 of the low-permeability layer within the model area vary from 0 to 50m, and values of vertical flow velocity w vary from -3×10^{-4} to $+2.2 \times 10^{-3}$ m/day.

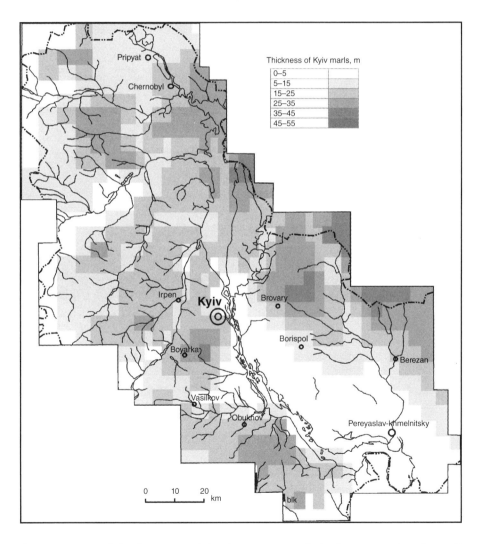

Figure 5.9 Map of the low-permeability layer (Kyiv marl) thickness m_0 taken from 3D hydrogeological model of the Kyiv region [*Rudenko et al.*, 1997; *Shestopalov et al.*, 1997].

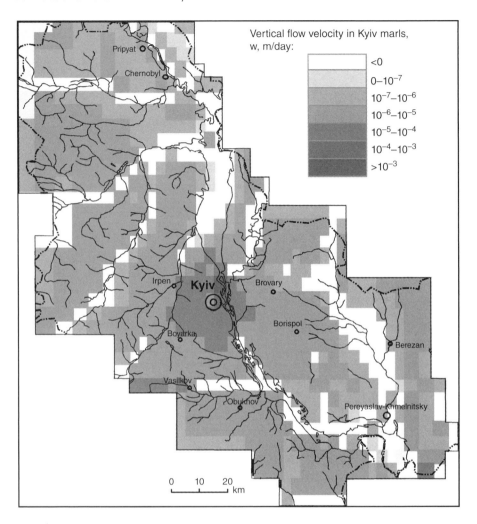

Figure 5.10 Map of vertical flow velocity w in the low-permeability layer = of Kyiv marls taken from a 3D hydrogeological model of the Kyiv region [*Rudenko et al.*, 1997; *Shestopalov et al.*, 1997].

White zones on the flow velocity map in Figure 5.10 correspond to areas with upward flow or negative values of w. The highest positive values of w (maximum downward flow) are observed in the area of the depression cone formed by the groundwater intakes of Cenomanian-Callovian (third from the surface) and Bajocean (fourth from the surface) aquifers operating for the water supply of Kyiv City.

Following the overlaying procedure for the two maps discussed above, the average-value zones for velocities w and thicknesses m_0 were determined. For these zones

Table 5.1 Evaluation gradations for Eocene aquifer protectability index $\varepsilon_2 = -\log$ $c_2(z_2^*, t^*)$ determined by total attenuation capacity of the saturated part of Quaternary aquifer and low-permeable Kyiv marl bed.

Vertical Flow Velocity Range, m/day	Aquifer Protectability Index ε_2 by Interval (Average Value) of Total Rock Thickness, m				
	0–20(15)	20–30(25)	30–40(35)	40–50(45)	50–60(55)
10^{-7}–10^{-5}	1	8	9	9	9
10^{-5}–10^{-4}	0	2	4	8	9
10^{-4}–10^{-3}	0	0	2	4	6
$>10^{-3}$	0	0	0	1	2

Numbers in parentheses mean the average depth for a given depth interval.

the relative concentration $c_2(z_2^*, t^*)$ of ^{137}Cs has been evaluated using the corresponding modeling plots (Figure 5.8) at depth z_2^* of the Eocene aquifer (second from the surface) for the forecast time period $t^* = 30$ years. This evaluation accounts also for the average thickness of the saturated part of the Upper Quaternary aquifer. After that, for every zone, the groundwater protectability index $\varepsilon_2 = -\log c_2(z_2^*, t^*)$ is evaluated. It characterizes the Eocene aquifer protectability determined by the total attenuation capacity of the saturated part of the upper Quaternary aquifer and the Kyiv marl low-permeable bed. Table 5.1 gives evaluation gradations of the averaged index ε_2 for corresponding zones of low-permeable layer thickness m_0 and vertical flow velocity w.

According to the gradations of Table 5.1, the map of the relative concentration $c_2(z_2^*, t^*)$ has been drawn reflecting the attenuation capacity of the low-permeable bed and the saturated part of the Quaternary aquifer. Onto this map, the previously drawn map of cover protectability of the Quaternary aquifer (Figure 5.5) has been "overlaid" through the summing of groundwater protectability indices [equation (4.9)]. As a result, the assessment map of cover Eocene aquifer protectability has been drawn. It accounts for the whole bed of Quaternary aquifer deposits and the attenuation capacity of the Kyiv marl low-permeable bed. The value zones of the aquifer protectability index ε are shown in the gray scale in Figure 5.11. Corresponding ranges of ^{137}Cs relative concentration in percents are given in the map legend below the gray-scale gradations.

The zones with a relatively low protectability index for the Eocene aquifer are associated mainly with areas of absence, weak occurrence or a disturbance (deep PFMZ) of the low-permeable Kyiv marl bed (see Figure 5.9), and areas of increased downward flow velocity (Figure 5.10). These zones are situated mainly in the middle and southeastern parts of the map adjacent to the Dnieper and the valleys of its largest tributaries (Pripyat, Uzh, Zdvizh, and Teterev).

Also characteristic are low-protected zones around the operating groundwater intakes of Kyiv City and the CEZ (Pripyat town), which form high downward flow velocities in the cover deposits of the assessed Eocene aquifer. These downward

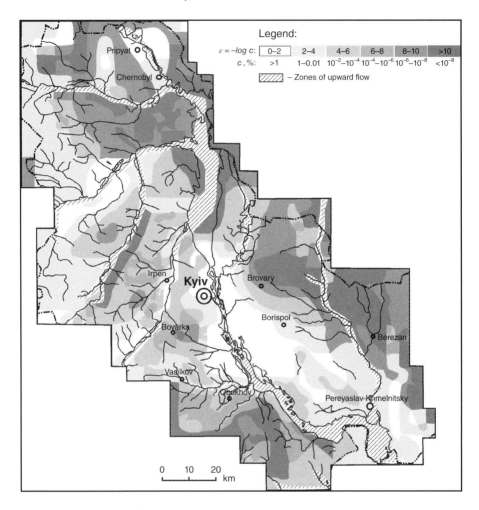

Figure 5.11 Map of the Eocene aquifer protectability from contamination by ^{137}Cs for the Kyiv region (within area of the 3D groundwater flow model). Forecast period 30 years.

flows increase even more in places of weakly occurring (by thickness) and disturbed low-permeable Kyiv marl beds. Dashed areas on the map show the groundwater discharge zones with upward flow in the described Eocene aquifer. These are narrow zones along the water courses of the Dnieper and its main tributaries and reservoirs. Zones of relatively high aquifer protectability correspond to watershed areas with a commonly occurring confining low-permeable bed and significant total thickness of the covering deposits.

By the protectability index ε, the corresponding relative concentration of the contaminant can be assessed in the map zon. It characterizes the total permeability of the upper aquifer and the confining bed:

$$c\left(z_2{}^*,t^*\right)=10^{-\varepsilon}. \tag{5.1}$$

Finally, this characteristic is considered a measure of intrinsic cover vulnerability of the Eocene aquifer.

Using the data of the upper Quaternary aquifer's contamination with Chernobyl-born ^{137}Cs, the vulnerability of the second (from the surface) Eocene aquifer to contamination with this radionuclide is estimated. The assessment is performed by overlaying the map of the Quaternary aquifer cover vulnerability to real surface contamination (Figure 5.6) onto the working map of the total attenuation capacity of the Quaternary aquifer saturated zone and the underlying Kyiv marl confining bed. The corresponding protectability index $\varepsilon_2 = -\log c_2(z_2{}^*,t^*)$ was used to draw the cover protectability map of the Eocene aquifer. As a result, the vulnerability map of the Eocene aquifer to real surface contamination with ^{137}Cs is obtained as shown in Figure 5.12. The map gradations correspond to zones of total groundwater contamination of the Eocene aquifer for the forecast period of 30 years in units of surface contamination density, kBq/m^2 (Figure 5.12, column A) and Ci/km^2 (column B). In the same way as for the upper Quaternary aquifer, by the obtained sum of contamination per unit area of the aquifer with account of its thickness, the average effective porosity $n = 0.1$, and average distribution coefficient $K_d = 10\,\mathrm{dm^3/kg}$ [*Shestopalov*, 2001], an approximate assessment is performed of the full groundwater vulnerability of the aquifer groundwater in units of ^{137}Cs concentration (mBq/dm^3; Figure 5.12, column C).

Similar to the vulnerability map of the upper groundwater, regions of highest groundwater vulnerability (3 mBq/dm^3 and higher) are observed in the CEZ. However, their areas here are restricted to sites of relatively low groundwater protectability (see Figure 5.11). The zones with a groundwater vulnerability of 0.03 mBq/dm^3 and higher in Kyiv conurbation and the Dnieper valley southeast of Kyiv can be associated with zones of absence or low thickness of the Kyiv marl confining bed, the presence of hydraulic "windows", and also the influence of Kyiv groundwater intakes in the deeper Cenomanian-Callovian and Bajocean aquifers. In the whole, the vulnerability of the confined Eocene aquifer is significantly lower as compared to the upper Quaternary aquifer.

It is worth mentioning that the maximum allowable concentration (MAC[1]) in drinking water for ^{137}Cs and ^{90}Sr according to standards of Ukraine is 1 Bq/dm^3.

---

[1]   The term MAC corresponds to the definition of the Maximum Contaminant Level (MCL) in the USA. U.S. EPA has established a MCL of 4 millirems per year for beta particle and photon radioactivity from man-made radionuclides in drinking water. The average concentration of strontium-90 that is assumed to yield 4 millirems per year is 8 picoCuries per liter (pCi/L), which equal to 0.3 Bq/l. (Reference: EPA Facts about strontium-90, see http://www.epa.gov/oerrpage/superfund/health/contaminants/radiation/pdfs/Strontium-90%20Fact%20Sheet%20final.pdf)

For sesium-137, the criteria is 200 pCi/L, which is equal to 7.4 Bq/l. (Reference: EPA Facts about Cesium-137, see: http://www.epa.gov/oerrpage/superfund/health/contaminants/radiation/pdfs/Cesium-137%20Fact%20Sheet%20final.pdf)

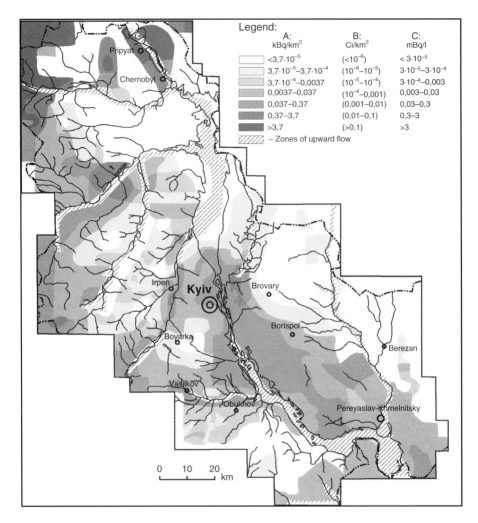

Figure 5.12 Vulnerability map of the Eocene aquifer to contamination with [137]Cs: (A, B) assessment of cover vulnerability of the aquifer in units of surface contamination density; (C) assessment of full groundwater vulnerability with account of aquifer capacity (in units of predicted concentration). Forecast period 30 years.

According to the World Health Organization, the MAC for emergency conditions for these nuclides is 10 Bq/dm^3.

Hence, according to the obtained assessments, even in conditions of high surface radioactive contamination in the CEZ, groundwater contamination of the aquifers with ^{137}Cs within most of the area remains below the MAC approved for water supply purposes.

6. SUMMARY

The results of field observations along with experimental and modeling studies of migration of Chernobyl-born radionuclides in the groundwater system within the CEZ and the Kyiv region clearly indicate that radioactive contaminants reached the groundwater through depression-related PFMZs. Depending on the surface depression density and groundwater depths, contaminant transport through the depressions could be from 3–30% to 80–90% and more of the total mass transport. The Chernobyl-born ^{137}Cs penetrates into the upper Quaternary aquifer within the entire area of the CEZ and the Kyiv region. The intensity of the ^{137}Cs migration varies over the area, depending on the upper groundwater protectability index. Zones of minimal protectability of the upper groundwater correspond to areas of high occurrence of depression-related PFMZs: northern and eastern parts of the study area, most areas of Polesye, sandur plains, and over floodplain terraces. The sites with a perfect protected upper groundwater aquifer are absent within the study area. The obtained map of the upper aquifer protectability for ^{137}Cs characterizes also the aquifer intrinsic vulnerability independent on the surface contamination distribution. The groundwater vulnerabilty can be considered as the inverse function of the protectability. In this sense the zones of lower aquifer protectability correspond to zones of higher aquifer vulnerability.

The vulnerability of the upper Quaternary aquifer and the vulnerability of the whole groundwater system to contamination with Chernobyl-born ^{137}Cs (Figure 5.5), as dependent on the surface contamination distribution, was determined based on the results of field monitoring along with numerical modeling of the study area after the Chernobyl catastrophe. The zones with the highest groundwater vulnerability to ^{137}Cs (by equivalent surface contamination density over 1 Ci/km^2 and by concentration in groundwater over 100 mBq/dm^3) correspond to areas with maximum surface contamination in the CEZ. The zones of high groundwater vulnerability correspond to the territories of Polesye and multiple areas of increased surface contamination. The areas of lowest-vulnerable and low-vulnerable groundwater correspond to relatively small watershed areas in the southern

Groundwater Vulnerability: Chernobyl Nuclear Disaster, Monograph Number 69.
Edited by Boris Faybishenko and Thomas Nicholson.

and southwestern parts, respectively, of the Kyiv region and to separate clean spots. The majority of the study area is characterized by average groundwater vulnerability with equivalent ^{137}Cs surface contamination densities from 0.01 to 0.1 Ci/km^2 and concentration in groundwater within 1–10 mBq/dm^3.

The assessment of vulnerability and protectability of the second (from the surface) Eocene aquifer was performed using the developed methodology with account of data of independent groundwater flow modeling with a four-layer hydrogeological model of the study area. The model provided the necessary data regarding vertical flow through the confining low-permeability bed of Kyiv marls. Consequently, the assessment was performed in such a way as to implicitly account for the transit flow into the underlying deeper aquifers (Cenomanian-Callovian and Bajocean), which are intensively exploited by the groundwater intakes for the water supply of Kyiv City. The obtained map of the Eocene aquifer protectability to contamination with Chernobyl-born [137]Cs (Figure 5.10) represents the areal distribution of the groundwater protectability index $\varepsilon = -\log c$ (where c is dimensionless contaminant concentration relative to the initial surface) with gradations of its value from 1 to 10 and corresponding gradations of relative concentration in percent. The map accounts for the total hydraulic and geochemical attenuation capacity of the upper Quaternary aquifer and the confining bed of Kyiv marls. Characteristic zones of relatively weak aquifer protectability correspond to areas of no or weak occurrence of the confining bed and its disturbances (PFMZs), which in turn are characterized by significant downward flow velocities. Depression cones from the operating water intakes of Kyiv City and Pripyat town (within CEZ) in the underlying Cenomanian-Callovian and Bajocean aquifers also have a noticeable influence on the formation of these weak protectability zones of the Eocene aquifer. These depression cones form increased downward flow components in the upper aquifer and the first confining bed.

Using the results of modeling of predicted groundwater contamination with ^{137}Cs of the upper Quaternary aquifer, the vulnerability map of the Eocene aquifer has been drawn in units of surface contamination density (kBq/m^2 and Ci/km^2; Figure 5.11). This was done by "overlying" the upper aquifer vulnerability map (Figure 5.5) onto the working map of the total attenuation capacity of the saturated part of the upper Quaternary aquifer and the Kyiv marl confining bed. The approximate assessment of the Eocene aquifer vulnerability to ^{137}Cs is performed in units of its predicted concentration in groundwater (mBq/dm^3). This was done by the recalculation of the total assessed cover contamination (in kBq/m^2) infiltrating into the aquifer in to the concentration in groundwater (mBq/dm^3) with account of the concentration decrease caused by sorption of contamination by aquifer rock calculated using data of aquifer thickness, porosity, rock density, and distribution coefficient K_d. The zones of highest Eocene aquifer vulnerability correspond to areas of highest surface contamination within the CEZ. However, their area is restricted within sites of relatively low aquifer protectability (Figure 5.10). On the whole, the assessed vulnerability of the confined Eocene aquifer is significantly lower as compared with that of the upper Quaternary aquifer.

The zones of Quaternary aquifer vulnerability show significantly higher groundwater vulnerability to contamination with ^{137}Cs as compared with previous assessments performed with no account of PFMZs [*Shestopalov*, 2003]. This proves once more the importance of PFMZ input into upper groundwater contamination by the infiltration of a contaminant from the surface. The results of the groundwater protectability assessment for Chernobyl-related radioactive contamination obtained with use of the developed methodology (with PFMZs taken into account) are in much better correlation with field data of groundwater contamination by ^{137}Cs [*Shestopalov*, 2001] than previously obtained with no account of PFMZs [*Belousova and Galaktionova*, 1994; *Belousova*, 2005] with no account of PFMZs.

Vyacheslav Shestopalov
Alexander Bohuslavsky
Volodymir Bublis
Boris Faybishenko
Thomas Nicholson

REFERENCES

Albinet, M., and J. Margat (1970), Cartographie de la vulnerabilité à la pollution des nappes d'eau souterraine, *Bull. BRGM 2nd Ser.*, *3*(4), 13–22.

Aller, L., T., Bennet J. H., Lehr R. J, Petty and G. Hackett (1987), DRASTIC: A standardized system for evaluating ground water pollution potential using hydrogeologic settings, EPA/600/2-87-036, U.S. Environmental Protection Agency, Ada, Washington, D.C.

Andersen, L. J., and E. Gosk (1987), Applicability of vulnerability maps, paper presented at the Intl. Conf. Vulnerability of Soil and Groundwater to Pollutants, *RIVM Proc. and Inf.*, *38*, 321–332.

Baker, R. S., and D. Hillel (1990), Laboratory tests of a theory of fingering during infiltration into layered soils, *Soil Sci. Soc. Am. J.*, *54*, 20–30.

Ball, D., A. MacDonald, B. Dochartaigh, M. del Rio, V. Fitzsimons, C. Auton, and A. Lilly. (2004), Development of a groundwater vulnerability screening methodology for the Water Framework Directive, Final report, Project WFD28, SNIFFER, www.sepa.org.uk/pdf/groundwater/tools/vulnerability_rgeport.pdf.

Baryakhtar, V. G., V. I., Kholosha and D. M. Grodzinsky (Eds.) (1997), Chernobyl Catastrophe. National Academy of Sciences of Ukraine, Ukrainian Ministry for the Population Protection from Chernobyl NPP Accident Consequences, Ministry of Health, Editorial House of Annual Issue "Export of Ukraine", Kyiv (In Russian).

Bear, J. (1972), *Dynamics of Fluids in Porous Media*, American Elsevier, New York.

Belousova, A. P. (2001), *Groundwater Quality: Present Approaches to the Assessment*, Nauka, Moscow, (in Russian).

Belousova, A. P. (2005), *Groundwater Resources and Their Security from Contamination in the River Dnieper Basin and Its Separate Regions: Territory of Russia*, URSS, Moscow. (in Russian).

Belousova, A. P., and Galaktionova O. V. (1994), On the methodology of assessing natural groundwater security from radioactive contamination, *J. Water Resources, Moscow*, *21*(3), 340–345 (in Russian).

Beven, K., and P.Germann (1982), Macropores and water flow in soils, *Water Resources Res.*, *18*(5), 1311–1325.

Bochever, F. M., and A. E. Oradovskaya (1972), *Hydrogeological Fundamentals of Groundwater Protection from Contaminants*, Nedra, Moscow (in Russian).

Groundwater Vulnerability: Chernobyl Nuclear Disaster, Monograph Number 69.
Edited by Boris Faybishenko and Thomas Nicholson.
© 2015 American Geophysical Union. Published 2015 by John Wiley & Sons, Inc.

Borzilov, V. A. (1989), Physical-mathematical modeling of processes determining runoff of long-lived radionuclides from watersheds in 30-km zone of Chernobyl NPP, *J. Meteorol. and Hydrol.*, *1*, 5–13 (in Russian).

Bouma, J. (1981), Soil morphology and preferential flow along macropores, *Agric. Water Manag.*, *3*, 235–250.

Burkart, M. R., D. W., Kolpin and D. E. James (1999), Assessing groundwater vulnerability to agrichemical contamination in the Midwest US, *Water Sci. and Technol.*, *39*(3), 103–112.

Carter, A. D., R. C. Palmer, and R. A. Monkhouse (1987), Mapping the vulnerability of groundwater to pollution from agricultural practice, particularly with respect to nitrate, *Atti Int. Conf. Vulnerab. of Soil and Groundwater to Pollutants, RIVM Proc. and Inf.*, *38*, 333–342.

Ciang, W. H., and W. Kinzelbach (2001), *3D Groundwater Modeling with PMWin*, Springer-Verlag Berlin Heidelberg.

Civita, M. (2008), The Italian "combined" approach in assessing and mapping the vulnerability of groundwater to contamination, in Zlatko Mikulič; Mišo Andjelov. *Proceedings of Invited Lectures of Symposium on Groundwater Flow and Transport Modelling*, Ljubljana, Sovenia, 28–31 January 2008, MOP - Agencija RS za okolje, Ljubljana, pp. 17–28.

Civita, M., and M. De Maio (2004), Assessing and mapping groundwater vulnerability to contamination: The Italian "combined" approach, *Geofis. Int.*, *43*(4), 513–532.

Daly, D., A. Dassargues, D. Drew, S. Dunne, N. Goldscheider, S. Neale, I. C. Popescu, and F. Zwahlen (2002), Main concepts of the European approach for (karst) groundwater vulnerability assessment and mapping, *Hydrogeol J.*, *10*, 340–345.

Deecke, W. (1906), Einige Beobachtungen am Sandstrande, Centralbl. fuer Mineral. Geol. Und Palaeont., Stuttgart. pp. 721–727.

Denny, S. C., D. M., Allen and J. M. Journeay (2007), DRASTIC-Fm: A modified vulnerability mapping method for structurally controlled aquifers in the southern Gulf Islands, British Columbia, Canada, *Hydrogeol. J.*, *15*(3), 483–493.

Doerfliger, N., P.-Y. Jeannin, and F. Zwahlehn (1999), Water vulnerability assessment in karst environments: A new method of defining protection areas using a multi-attribute approach and GIS tools (EPIK method), *Environ. Geol.*, *39*(2), 165–176.

Engel, B., K. Navulur, B. Cooper, and L. Hahn (1996), Estimating groundwater vulnerability to nonpoint source pollution from nitrates and pesticides on a regional scale, HydroGIS96: Application of Geographic Information Systems in Hydrology and Water Resources Management (Proceedings of the Vienna Conference, April, 1996), IAHS Publ. 235.

Engelen, G. B. (1985), Vulnerability and restoration aspects of groundwater systems in unconsolidated terrains in the Netherlands, Atti 18 Cong. I.A.H., pp. 64–69.

Evans, T. A., and D. R. Maidment (1995), A spatial and statistical assessment of the vulnerability of Texas groundwater to nitrate contamination, Center for Research in Water Resources, Bureau of Eng. Res., Univ. of Texas at Austin, J. J. Pickle Res. Campus, Austin, http://civil.ce.utexas.edu/centers/crwr/reports/online.html.

Faybishenko, B., C., Doughty M., Steiger J. C. S., Long T. R., Wood J. S., Jacobsen J. Lore, and P. T. Zawislanski (2000), Conceptual model of the geometry and physics of water flow in a fractured basalt vadose zone, *Water Resources Res.*, *36*(12), 3499–3520.

Faybishenko, B., P. A. Witherspoon, and J. Gale (Eds.) (2005), *Dynamics of Fluids and Transport in Fractured Rock*, Geophysical Monograph Ser. 162, Am. Geophys. Union, Washington D.C.

Fetter C. W. (2000), *Applied Hydrogeology*, Prentice Hall, Englewood Cliffs, N.J.

Foster, S. S. D. (1987), Fundamental concepts in aquifer vulnerability, pollution risk and protection strategy, Atti Int. Conf.Vulnerab. of Soil and Groundw. to Pollutants, RIVM Proc. and Inf. 38, pp. 69–86.

Foster, S. S. D., and R. Hirata (1988), Groundwater pollution risk assessment: A methodology using available data, Pan Amer. Cent. for Sanit. Engin. and Envir. Scienc. (CEPIS), Lima.

Freeze, R. A., and J. A. Cherry (1979), *Groundwater*, Prentice-Hall, Englewood Cliffs, N.J.

Fried, J. (1975), *Groundwater Pollution*, Elsevier, Amsterdam, New York.

Gees, R. A., and A. K. Lyall (1969), Erosion sand columns in dune sand, Cape Sable Island, Nova Scotia, Canada, *Can. J. Earth Sci.*, *6*, 344–347.

Gerke, H. H., P. Germann, and J. Nieber (2010), Preferential and unstable flow: From the pore to the catchment scale, *Vadose Zone J.*, No 9, pp. 207–212.

Gladkiy, A. V., I. I. Lyashko, and G. E. Mistetskiy (1981), *Algorithmization and Numeric Calculation of Filtration Schemes*, Vyscha Shkola, Kiev (in Russian).

Glass, R. J., T. S. Steenhuis, and J.-Y. Parlange (1989), Mechanism for finger persistence in homogeneous, unsaturated, porous media: Theory and verification, *Soil Sci.*, *148*, 60–70.

Gogu, R. C., and A. Dassargues (2000), Current trends and future challenges in groundwater vulnerability assessment using overly and index methods, *Environ. Geol.*, *39*(6), 549–559.

Goldberg, V. M. (1983), Natural and technogenic factors of groundwater protectability. Moscow Soc. of Nature Investigators Bull. No. 2, pp. 103–110 (in Russian).

Goldberg, V. M. (1987), *Interrelation of Groundwater Contamination and Natural Environment*, Gidrometeoizdat, Leningrad (in Russian).

Goldscheider, N. (2005), Karst groundwater vulnerability mapping: Application of a new method in the Swabian Alb, Germany, *Hydrogeol. J.*, *13*(4), 555–564.

Goman, A. V. (2005), Complex approach to hydrogeological protectability of subsurface geosphere of Astrakhan gas-condensate deposit. In: *Proceedings of the International Conference "Fundamental Problems of Oil-Gas Hydrogeology" devoted to 80th Jubilee of Prof. A. A. Kartsev*, Moscow, Institute of Oil and Gas Problems, Russian Academy of Sciences, 25–27 October 2005 (in Russian).

Goman, A. V. (2007), Hydrogeochemical protectability of atom-hydro-lithosphere in conditions of oil-gas fields recovery, Problems of geology and minerals production, in *Proceedings of All-Russia Scientific Conference*, 22–24 October 2007, pp. 260–262, Tomsk Polytech. Univ. Tomsk, Russia. (in Russian).

Greenland, D. J. (1977), Soil drainage by intensive arable cultivation: Temporary or permanent? *Phil. Trans. Roy. Soc. (Lond.)*, *B281*, 193–208.

Gripp, K. (1961), Ueber Werden und Vergehen von Barchanen an der Nordsee-Kueste Schleswig-Holsteins, *Zeitsch. fuer Geomorphologie*, Neue Folge, Bd. 5:24–36, Berlin, Germany.

Groundwater Resources of Southern Wisconsin (2002), Southeastern Wisconsin Regional Planning Commission, Wisconsin Geol. and Natural History Surv., Wisconsin Dept. of Natural Resources, Tech. Rept. 37, June 2002, www.sewrpc.org/publications/techrep/tr-037_groundwater_resources.pdf.

Grove, D. B., and K. G. Stollenwerk (1984), Computer model of one-dimensional equilibrium-controlled sorption processes, Water-Resources Investigations Rept. 84-4059, U.S. Geol. Surv., Reston, VA.

Gurdak, J. J., M. A. Walvoord, and P. B. McMahon (2008), Susceptibility to enhanced chemical migration from depression-focused preferential flow, High Plains aquifer, *Vadose Zone J.*, *7*(4), 1172–1184.

Haustov, A. P. (2007), *Stability of Subsurface Hydrosphere and Foundations of Ecological Regulation*, GEOS, Moscow (in Russian).

Heath, R. C. (1984), Ground-water regions of the United States, Water-Supply Paper 2242, U.S. Geol. Surv., Reston, VA.

Helling, C. S., and T. J. Gish (1991), Physical and chemical processes affecting preferential flow, in *Preferential Flow. Proceedings of the National Symposium*, 16–17 December, 1991, Chicago, Ill, edited by N. J. Gish and A. Shirmohammadi. *Am. Soc. of Agric. Eng.* Chicago, Ill.

Hill, S. (1952), Channeling in packed columns, *Chem. Eng. Sci.*, *1*, 247–253.

Hillel, D., and R. S. Baker (1988), A descriptive theory of fingering during infiltration into layered soils, *Soil Sci.*, *146*, 51–56.

Hoelting, B., T., Haertle K. H., Hohberger K. H., Nachtigall E., Villinger W., Weinzierl J. P. Wrobel (1995), Konzept zur Ermittlung der Schutzfunktion der Grundwasser ueber-deckung, *Geol Jahrb*, *C63*, 5–24.

Josopait, V., and B. Schwerdtfeger (1979), Geowissenschaftliche Karte des Naturraumpotentials von Niedersachsen und Bremen, CC 3110 Bremerhaven Grundwasser, 1:200000, Niedersachsischen Landesamt fur Bodenforshung, Hanover.

Jury, W. A., and K. Roth (1990), *Transfer Functions and Solute Movement through Soil: Theory and Applications*, Birkhaeuser Verlag.

Kholosha, V. I., N. I. Proskura, Yu. A. Ivanov, S. V. Kazakov, and A. M. Arkhipov (1999), Radiation and ecological significance of natural and technogenic objects of the exclusion zone, *Bull. Ecol. State of the Exclusion Zone and Zone of Mandatory Depopulation*, No. 13, 1999, Chernobyl.

Kraynov, S. R., and V. M. Schvets (1987), *Geochemistry of Groundwater Used for Potable and Industrial Needs*, Nedra, Moscow (in Russian).

Kraynov, S. R., B. N. Ryzhenko, and V. M. Schvets (2004), *Groundwater Geochemistry: Theoretical, Applied, and Ecological Aspects*, Nauka, Moscow (in Russian).

Kung, K-J. S. (1990), Preferential flow in a sandy vadose zone. 1. Field observation. 2. Mechanism and implications, *Geoderma*, *46*, 51–71.

Landon, J. R. (Ed.) (1984), *Booker Tropical Soil Manual*, Booker Agric. Int. Ltd., London.

Lawes, J. B., J. H. Gilbert, and R.Warington (1882), On the amount and composition of the rain and drainage water collected at Rothamsted, Williams Clowes and Sons, London. Originally published in *J. Royal Agr. Soc. England*, *17*(1881), 241–279, 311–350; *18*(1882), 1–71.

Ligget, J. E., and S. Talwar (2009), Groundwater vulnerability and integrated water resource management, *Streamline Watershed Management Bull.*, *13*(1), 18–29.

Lissey, A. (1971), Depression-focused transient groundwater flow patterns in Manitoba, Special Paper 9:333–341, Geological Association of Canada.

Loague, K., and D. Corwin (1998), Regional-scale assessment of non-point source groundwater contamination, *Hydrol. Process.*, *12*, 957–965.

Loague, K., R. H., Abrams, S. N., Davis, A., Nguyen, and I. T. Stewart (1998), A case study simulation of DBCP groundwater contamination in Fresno County, California: 2. Transport in the saturated subsurface, *J. Contaminant Hydrol.*, *29*, 137–163.

Lukner, L., and V. M. Shestakov (1986), *Modeling Groundwater Migration*, Nedra, Moscow (in Russian).

Magiera, P. (2000), Methoden zur Abschaetzung der Verschmutzungsempfindlichkeit des Grundwassers, *Grundwasser*, *3*, 103–114.

Maloszevski, P., and A. Zuber (1996), Lumped parameter models for the interpretation of environmental tracer data, IAEA TEC DOC – 910, Manual on mathematical models in isotope hydrogeology, Vienna, pp. 9–58.

Marcolongo, B., and L. Pretto (1987), *Vulnerabilita degli acquiferi nella pianura a nord di Vicenza*, Pubbl. GNDCI-CNR n. 28, Ed. Grafiche Erredieci, Padova.

Margane, A., M. Hobler, and A. Sabah (1999), Mapping of groundwater vulnerability and hazards to groundwater in the Ibrid area in N. Jordan, *Z. Angew. Geol.*, *45*, 4.

Margat, J. (1968), Vulnerabilite des nappes d'eau souterraine a la pollution, BRGMPublication 68 SGL 198 HYD, Orleans

McCoy, J. (2004), ArcGIS 9, Geoprocessing in ArcGIS, ESRI, Redlands, CA.

McDonald, M. C., and A. W. Harbaugh (1988), MODFLOW, A modular three-dimensional finite difference ground-water flow model, Open-file report 83–875, Chapter A1, *U. S. Geol. Surv.*, Reston, VA.

Mickhevich, G. S. (2011), Geoecological assessment of groundwater natural vulnerability to contamination (on the example of the upper inter-moraine aquifer system in Kaliningrad Oblast), Abstract of Candidate of Science Thesis, Kaliningrad, www.kantiana.ru/postgraduate/announce/avt_mihneviy.doc (in Russian).

Mironenko, V. A., and V. G. Rumynin (1990), Assessment of protective properties of aeration zone (as applied to groundwater contamination), *J. Eng. Geol.*, No 2, pp. 3–18 (in Russian).

Mironenko, V. A., and V. G. Rumynin (1999), *Problems of Hydrogeoecology*. vol. *3*, State Univ. of Mining, Moscow (in Russian).

Mironenko, V. A., E. V. Molskiy, and V. G. Rumynin (1988), *Studying Groundwater Contamination in Mining Regions*, Nedra, Leningrad (in Russian).

National Research Council (NRC) (1993a), National Academy Press Report, *Ground Water Vulnerability Assessment. Contamination Potential Under Conditions of Uncertainty*, Committee on Techniques for Assessing Ground Water Vulnerability Water Science and Technology Board, Commission on Geosciences, Environment, and Resources, Nat. Res. Council, Natl. Acad. Press, Washington, D.C.

National Research Council (NRC) (1993b), *Ground Water Vulnerability Assessment: Contamination Potential Under Conditions of Uncertainty*, Natl. Acad. Press, Washington DC, http://www.nap.edu/catalog/2050.html.

National Research Council (NRC) (2004), *Groundwater Fluxes Across Interfaces*. Natl. Acad. Press, Washington, D.C., http://www.nap.edu/catalog/10891.htm.

Nieber, J. L. (1996), Modeling finger development and persistence in initially dry porous media, *Geoderma*, *70*, 209–229.

Nieber, J. L. (2001), The relation of preferential flow to water quality, and its theoretical and experimental quantification, in *Preferential Flow: Water Management and Chemical Transport in the Environment, Proceedings of the 2nd International Symposium*, 3–5 January 2001, Honolulu, Hawaii, USA, Eds. D.D. Bosch and K.W. King, pp. 1–9, Am. Soc. of Agric. Eng. St. Joseph, Michigan.

Nieber, J. L., C. A. S. Tosomeen, and B. N. Wilson (1993), Stochastic-mechanistic model of depression-focused recharge, in Y. Eckstein and A. Zaporozec (Eds.), *Hydrologic Investigation, Evaluation, and Ground Water Modeling, Proceedings of Industrial and Agricultural Impacts on the Hydrologic Environment, The Second USA/CIS Joint Conference on Environmental Hydrology and Hydrogeology*, Water Environment Federation, Washington D.C. pp. 207–234.

Olmer, M., and B. Rezac (1974), Methodical principles of maps for protection of ground water in Bohemia and Moravia scale 1:200000, Mem. I.A.H. 10, 1, pp. 105–107.

Ostry, R. C., R. E. J., Leech A. J. Cooper, and E. H. Rannie (1987), Assessing the susceptibility of ground water supplies to non-point source agricultural contamination in

South Ontario, Atti Int. Conf.Vulnerab. of Soil and Groundw. to Pollutants, RIWM Atti and Inf. 38, pp. 437–445.

Palmer, R. C. (1988), Groundwater vulnerability Map Severn Trent Water, Soil Survey and Land Res. Cent. 8 p. 7 Carte.

Palmquist, R., and L. V. A. Sendlein (1975), The configuration of contamination enclaves from refuse disposal sites on floodplains, *GroundWater, 13*(2), 167–181.

Parlange, J.-Y., T. S. Steenhuis, R. J. Glass, T. L. Richards, N. B. Pickering, W. J. Waltman, N. O. Bailey, M. S. Andreini, and J. A. Throop (1988), The flow of pesticides through preferential paths in soils, *New York's Food & Life Sci. Quarterly, 18*(1, 2), 20–23 (Cornell University, Ithaca, N.Y.).

Pashkovskiy, I. S. (2002), Principles of assessing groundwater security from contamination, *J. Present problems of hydrogeology and hydromechanics*, Saint-Petersburg Univ., pp. 122–131 (in Russian).

Perelman, A. I. (1961), *Geochemistry of Landscape*, Geographgiz, Moscow (in Russian).

Phillip, J. R. (1975), Stability analysis of infiltration, *Soil Sci. Soc. Amer. Proc., 39*, 1042–1049.

Pityeva, K. E. (1999), *Hydrogeological Studies in Regions of Oil and Gas Deposits*, Nedra Moscow (in Russian).

Pityeva, K. E., A. V. Goman, and A. O. Serebryakov (2006), Groundwater geochemistry in oil-gas deposits development, Astrakhan Univ., Astrakhan (in Russian).

Polubarinova-Kochina, P. Ya. (1977), *Theory of Groundwater Movement*, Science, Moscow (in Russian).

Polyakov, V. A., and E. V. Golubkova (2007), Assessment of groundwater security using data of isotope hydrogeochemical research, *J. Earth's Interiors Prospecting and Protection*, No 5, pp. 48–52. Moscow. (in Russian).

Prazak, J., M. Sir, F. Kubik, J. Tywoniak, and C. Zarcone (1992), Oscillation phenomena in gravity-driven drainage in coarse porous media, *Water Resour. Res., 28*, 1849–1855.

Raats, P. A. C. (1973), Unstable wetting fronts in uniform and nonuniform soils, *Soil Sci. Soc. Amer. Proc. 37*, 681–685.

Rogachevskaya, L. M. (2002), Regional assessment of groundwater vulnerability of the Eastern part of Dnieper Artesian Basin to radionuclide contamination, Candidate of Geological Sciences' Thesis, Inst. of Water Problems, Moscow (in Russian).

Rogovskaya, N. V. (1976), Map of natural groundwater protectability from pollution. *J. Nature*, No 3, pp. 57–76. Moscow. (in Russian).

Rosen, L. (1994), A study of the DRASTIC methodology with emphasis on Swedish conditions, *Ground Water, 32*, 278–285.

Rudenko, Yu. F., V. M., Shestopalov A. S. Boguslavsky, and B. D. Stetsenko (1997), Groundwater use management based on permanent action models, *Proceedings of the XXVII IAH Congress on Groundwater in the Urban Environment, Nottingham, UK, 21–27 September, 1997, vol. 1, Problems, Processes and Management*, Ed.: J. Chilton. pp. 653–658, Balkema, Rotterdam Brookfield.

Rumynin V. G. (Ed.) (2003), Assessment of the influence of a nuclear industry complex on groundwater and related natural objects (town Sosnovy Bor, Leningrad region), Saint-Petersburg Univ. (in Russian).

Rundquist, D. C., A. I. Peters, L. Di, D. A. Rodekohr, R. L. Ehrman, and G. Murray, (1991), Statewide groundwater-vulnerability assessment in Nebraska using the DRASTIC/GIS model, *Geocarto Int. 2*, 51–58.

Schmidt, R. R. (1987), Groundwater contamination susceptibility in Wisconsin, Wisc. Groundw. Manag. Plan Rep. 5, 27 p.

Schnoor, J. L. (Ed.) (1992), *Fate of Pesticides and Chemicals in the Environment*, Wiley, New York.

Shestakov, V. M. (2003), Accouting for geological heterogeneity: A key problem of hydrogeodynamics, *J. Tribune of Moscow Univ.*, *4*(1) 25–27 (in Russian).

Shestopalov, V. M. (1979), *Groundwater Natural Resouces Formation in Platform Structures of Ukraine*, Naukova Dumka, Kiev (in Russian).

Shestopalov, V. M. (1981), *Natural Groundwater Resources of Platform Artesian Basins of Ukraine*, Naukova Dumka, Kiev (in Russian).

Shestopalov, V. M. (Ed.) (1988), *Water Exchange in Hydrogeological Structures of Ukraine: Methods of Water Exchange Study*, Naukova Dumka, Kiev (in Russian).

Shestopalov, V. M. (Ed.) (2001), Water exchange in hydrogeological structures of Ukraine: Water exchange in hydrogeological structures and Chernobyl disaster, Inst. of Geol. Sci., Radio-Environmental Center, Kyiv, Parts 1, 2 (in Russian).

Shestopalov, V. M. (Ed.) (2002), *Chernobyl Disaster and Groundwater*, Balkema, Lisse, Abingdon/Exton(Pa), Tokyo.

Shestopalov, V. M. (Ed.) (2003), Assessment of natural groundwater protectability from contaminants for Ukrainian area of Dnieper basin, Report on Research contract № BYE/00/001-01, October 28. Radio-Environmental Center NAS of Ukraine, Kyiv (in Russian).

Shestopalov, V. M., V. V. Gudzenko, Y. F. Rudenko, and A. S. Boguslavskij (1992), *Combined Analysis, Modelling and Forecast of Longterm Underground Water Contamination Inside the Chernobyl Fallout Influenced Zone, Hydrological Impact of Nuclear Power Plant Systems*, International Hydrological Programme, UNESCO Chernobyl Programme, Paris.

Shestopalov, V. M., A. S. Bohuslavsky, V. N. Bublias, V. V. Goudzenko, I. P. Onyschenko, and Yu. F. Rudenko (1996), Studying migration of Chernobyl-born radionuclides in groundwater used for drinking water supply of Kyiv city, *J. Chem. technol. water, Kyiv*, *18*(2), 120–127 (in Russian).

Shestopalov, V. M., V. V. Goudzenko, Yu. F. Rudenko, V. N. Bublias, and A. S. Boguslavsky (1997), Assessment and forecast of groundwater and rock contamination within the Kyiv industrial agglomeration influenced by Chernobyl fallout, in *Proceedings of the XXVII IAH Congress on Groundwater in the Urban Environment. Nottingham, UK, 21–27 September 1997, vol. 1, Problems, Processes and Management*, Ed.: J. Chilton. pp. 171–174, Balkema, Rotterdam, Brookfield.

Shestopalov, V. M., and V. N. Bublias (2000), Zones of intensive migration of radionuclides into geological environment of Chernobyl Exclusion Zone, *Bull. of Ecol. State Exclusion Zone and Zone of Mandatory Depopulation*, No 16, September 2000, "Chernobylinform" Agency, pp. 9–12, Kyiv.

Shestopalov, V. M., S. T., Zvolskiy V. M. Bublias, and V. V. Kulik (2002), Determination of infiltration characteristics in rocks of aeration zone by chlorine indicator, *J. Repts. Acad. Sci. Ukraine*, No 9, pp. 130–136, Kyiv. (in Ukrainian).

Shestopalov, V. M., Yu. F., Rudenko A. S. Bohuslavsky, and V. N. Bublias (2006), Chernobyl-born radionuclides: Aquifers protectability with respect to preferential flow zones, in *Applied Hydrogeophysics*, edited by H. Vereecken, pp. 341–376, Springer, Netherlands.

Shestopalov, V. M., A. B., Klimchuk S. V. Tokarev, and G. N. Amelichev (2009), Groundwater vulnerability assessment of regions of open karst (on example of the Ai-Petri massif, Crimea), *Speleol. Carstol., 2009*(2), 11–29 (in Russian).

Shuford, J. W., D. D. Fritton, and D. E. Baker (1977), Nitrate nitrogen and chloride movement through undisturbed field soil, *J. Env. Qual. 6*, 255–259.

Sililo, O. T. N., J. E. Conrad, T. E. Doehse, G. Tredoux, and M. H. du Plessis (2001), A procedure for deriving qualitative contaminant attenuation maps from land type data, *J. Hydrol.*, *241*, 104–109.

Singh, P., and R. S. Kanwar (1991), Preferential solute transport through macropores in large undisturbed saturated columns, *J. Environ. Qual.*, *20*, 295–300.

Sinreich, M., F. Cornation, and F. Zwahlen (2007), Evaluation of reactive transport parameters to assess specific vulnerability in karst systems, in *Groundwater Vulnerability Assessment and Mapping. Selected Papers from the Groundwater Vulnerability Assessment and Mapping International Conference, Ustron, Poland, 2004*, edited by A. J. Witkowski et al., pp. 38–48, Taylor & Francis/Balkema, Netherlands.

Sotornikova, R., and J. Vrba (1987), Some remarks on the concept of vulnerability maps, Atti Int. Conf.Vulnerab. of Soil and Groundw. to Pollutants, RIVM Proc. and Inf. 38, pp. 471–475.

Tovar, M., and R. Rodriguez (2004), Vulnerability assessment of aquifers in an urban-rural environment and territorial ordering in Leon, Mexico, *Geofis. Int.*, *43*(4), 603–609.

Van Genuchten, M. Th. (1980), A close-form equation for predicting the hydraulic conductivity of unsaturated soils, *Soil Sci. Soc. Am. J.*, *44*, 892–898.

Van Stempvoort, D., L. Ewert, and L. Wassenaar (1995), A method for groundwater protection mapping in the Praire Province of Canada, PPWB Report 114, Nat. Hydrogeol. Res. Inst., Saskatoon, Saskatchevan, Canada.

Villumsen, A., O. S. Jacobsen, and C. Sonderskov (1983), Mapping the vulnerability of ground water reservoirs with regard to surface pollution, Danm. Geol. Unders. Arbog 1982, pp. 17–38, 2 Tavole.

Vlaicu, M., and C-M. Munteanu (2008), Karst groundwaters vulnerablity assessment methods, Trav. Inst. Speol. "Emile Rakovitza," t. XLVII, pp. 107–118, Bucarest, www.speotravaux.iser.ro/08/art06.pdf.

Vogel, T., R. Zhang, M. Th. van Genuchten, and K. Huang (1995), HYDRUS, one-dimensional variably saturated flow and transport model, including hysteresis and root water uptake, Version 3.4, Res. Rept., U.S. Salinity Lab., USDA-ARS, Riverside, CA.

Von Hoyer, M., and B. Söfner (1998), Groundwater vulnerability mapping in carbonate (karst) areas of Germany, BGR Hannover, Archive Nr. 117 854, unpubl. report for EC-project COST Action 620.

Vrana, M. (1968), Ochrana prostyc podzemnich vod v Cechach a na Morava.Vysvetlivsky k mape 1:500000, *Wat. Res. Plan Cent.*, Praha (in Ceco).

Vrana, M. (1984), Methodology for construction of groundwater protection maps, Lecture for UNESCO/UNEP Proj. PLCE3/29, Moscow, September 1981, in *Hydrogeological Principles of Groundwater Protection*, vol. 1, E. A. Kazlovsky Edit. and Chief, pp. 147–149. UNESCO/UNEP, Moscow.

Vrba, J., and A. Zaporozec (Eds.) (1994), *Guidebook on Mapping Groundwater Vulnerability: International Contributions to Hydrogeology*, vol. *16*. Int. Assoc. of Hydrogeol., Heise, Hanover.

Vsevolzhskiy, V. A. (1983), *Underground Runoff and Water Balance of Platform Structures*, Nedra, Moscow (in Russian).

Zaporozec, A. (Ed.) (1985), *Groundwater Protection Principles and Alternatives for Rock County, Wisconsin*, Wisconsin Geological and Natural History Survey, Special Report 8, WI SR 8.

Zaporozec, A. (Ed.) (2002), *Groundwater Contamination Inventory*, IHP-VI, Series on groundwater No. 2, UNESCO, Paris.

Zektser I. S. (2001), *Groundwater as a Component of Environment*. Nauchny Mir, Moscow (in Russian).

Zektser, I. S. (2007), *Groundwater of the World: Resources, Use, Forecasts*, Nauka, Moscow (in Russian).

Zektser, I. S., O. A. Karimova, G. Buguoli, and M. Bucci (2004), Regional assessment of fresh groundwater vulnerability: Methodological aspects and practical application, *J. Water Resources*, *31*(6), 645–650 (in Russian).

Zhang, R., J. D. Hamerlinck, S. P. Gloss, and L. Munn (1996), Determination of non-point-source pollution using GIS and numerical models, *J. Environ. Quality*, *25*(3), 411–418.

Zheng, C. (1990), MT3D. A modular three-dimensional transport model for simulation of advection, dispersion and chemical reaction of contaminants in groundwater systems, S.S. Papadopulos & Assoc., Rockville, Md., prepared for the U.S. EPA Robert S. Kerr Environmental Research Laboratory, Ada, Okla., 17, October 1990.

Zhou, H., Guoli, W., and Qing, Y. (1999), A multi-objective fuzzy pattern recognition model for assessing groundwater vulnerability based on the DRASTIC system, Special Issue: Barriers to Sustainable Management of Water Quantity and Quality, *J. Hydrol. Sci.*, *44*(4), 611–618.

Zhou, J., G., Li F., Liu Y. Wang, and X. Guo (2010), DRAV model and its application in assessing groundwater vulnerability in arid area: A case study of pore phreatic water in Tarim Basin, Xinjiang, Northwest China, *Environ. Earth Sci.*, *60*(5), 1055–1063.

Zwahlen, F. (Ed.) (2004), Vulnerability and risk mapping for the protection of carbonate (karst) aquifers, final report COST Action 620, European Commission, Directorate-General for Research, EUR 20912, Luxemburg.

INDEX

Groundwater Vulnerability: Chernobyl Nuclear Disaster, Monograph Number 69.
Edited by Boris Faybishenko and Thomas Nicholson.
© 2015 American Geophysical Union. Published 2015 by John Wiley & Sons, Inc.